D0241881

03374420 - ITEM

BICARBONATE OF SODA

A Very Versatile Natural Substance

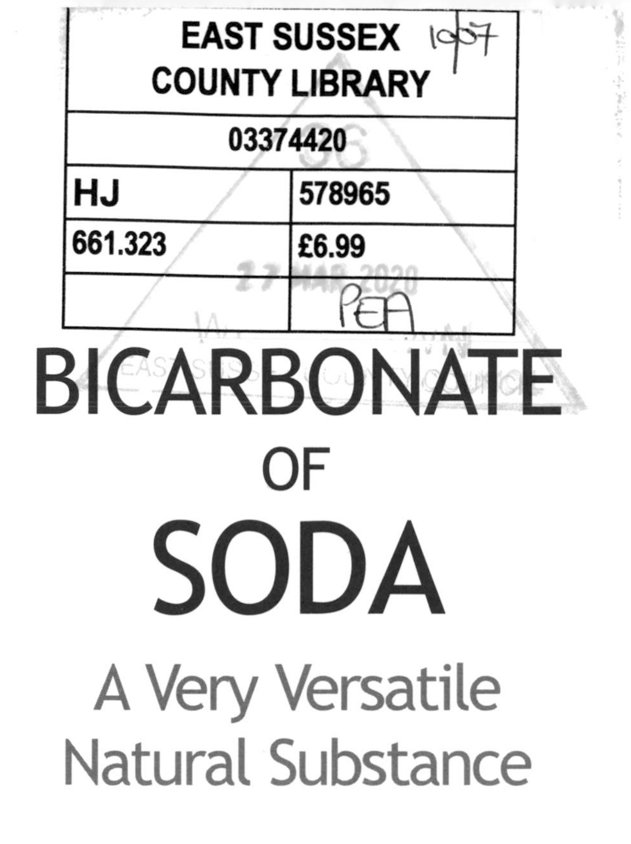

BICARBONATE OF SODA

A Very Versatile Natural Substance

Margaret Briggs

This edition published 2007
by Black & White Publishing Ltd
99 Giles Street, Edinburgh EH6 6BZ

ISBN 13: 978-1-84502-163-4
ISBN 10: 1-84502-163-0

1 3 5 7 9 10 8 6 4 2 07 08 09 10

Illustrations by Tegan Sharrard
Cover design by Omnipress Ltd

Printed in Dubai

ABOUT THE AUTHOR

Margaret Briggs was a teacher for 30 years, working in Kent, Germany, North Yorkshire and Sussex.

Since leaving teaching she has had more time for gardening and cooking and has embarked on a second career as a freelance writer, researcher and editor, alongside her writer husband, Lol. Six years ago the couple bought a dilapidated house in south-west France. The house is now restored and Margaret and Lol divide their time between Sussex and the Gironde, with two contrasting gardens to develop.

Margaret has written four other books in this series, *Vinegar – 1001 Practical Uses*, *Gardening Hints and Tips*, *Porridge – Oats and their Many Uses* and *Honey – and its many health benefits*.

CONTENTS

Introduction

A BOOK ABOUT BICARBONATE OF SODA

Do you ever lie awake at night, wondering about the difference between bicarbonate of soda, baking soda, baking powder and sodium carbonate? If so, this is the book for you; after, of course, you have consulted a specialist in sleeping disorders!

When I was approached to write this book I thought at first that it might be a bit of a joke, or at best a complete non-starter. My friends and family already wonder at my capacity for astounding and amusing them with titles such as *Vinegar, 1001 Practical Uses* or the health benefits of eating porridge or honey, but none of them expected *Bicarbonate of Soda* to be the sequel to *Honey*!

Having looked into the subject, however, and remembering how clued-up I now am on vinegar as a household cleaner, a vital kitchen aid and the bearer of many health benefits enjoyed by millions of people across the world, I could hear more than a faint bell ringing, reminding me of the number of times bicarbonate of soda comes up in recipes and remedies. There were bound to be rich, untapped seams of information and ideas out there on bicarb of soda, weren't there?

Did you know that bicarbonate of soda is a valuable commodity in the world of glass and soap manufacture? I'm sure you were already aware that the Ancient Egyptians used it to make their mummies smell better and stay fresh for longer? No wonder there are nowadays so many applications of bicarbonate of soda that can achieve acceptable results for half the cost of modern cosmetics and without unleashing even more harmful substances on our already damaged planet.

If only I'd done chemistry at school beyond the age of 14, I might understand all those chemical symbols – but I like a challenge. I spent the best part of 30 years explaining things to primary school pupils in easy terms and even finally enjoyed teaching science, when understanding reversible and irreversible changes meant messing about with bicarbonate of soda and vinegar to make volcanoes, so here goes.

If you like trivia, or take part in quizzes, some of it might come in useful. You may not like the chemical composition stuff, but you might find the fun science a welcome distraction, even if you haven't got the excuse of children or grandchildren to entertain. Some of us never grow up!

What is Bicarbonate of Soda?

THE CHEMISTRY STUFF

Bicarbonate of soda, a white chemical compound, has also been known as:

- sodium bicarbonate
- saleratus
- baking soda
- bread soda
- sodium hydrogen carbonate

For the more scientifically minded, the chemical formula is:

NaHCO³

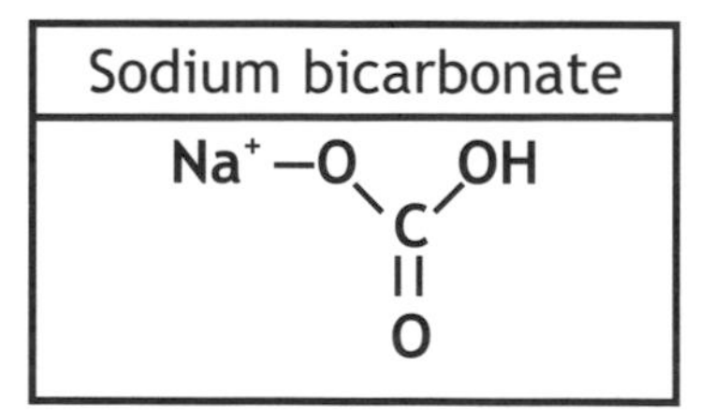

These symbols show you the constituent parts of this chemical compound. Sodium Bicarbonate is made from the elements carbon (C), hydrogen (H), oxygen (O) and sodium (Na). It can be reproduced by the Solvay process which involves the reaction of sodium chloride, ammonia and carbon dioxide and produces over 100,000 tons per year. Commercial quantities of bicarbonate of soda are also produced when soda ash, mined in the form of the ore trona, is dissolved in water and treated with carbon dioxide.

As a solid it is not totally soluble in water. It is crystalline in appearance but often looks like a powder. The Na in the chemical formula comes from natron, a substance known to be used as far back as Ancient Egyptian times, and found in many mineral springs. It has a taste resembling that of sodium carbonate and is alkaline in terms of pH value (8.3). Bicarbonate of soda is considered to be relatively safe to use, without harming the skin, but ingesting large amounts should be avoided.

Sodium is a soft, silver-white metallic element, one of the alkali metals, found in soda, salt and other compounds.

To avoid confusion, the following are not the same as bicarbonate of soda, although they have similar names and some similar constituents:

- Baking powder, which contains bicarbonate of soda
- Sodium chloride, or common salt
- Sodium hydroxide or caustic soda
- Sodium carbonate, known as soda ash, sodium decahydrate and washing soda

These all have wide-ranging applications and effects. You can find out more about all of these sodium-related substances in the glossary.

SODIUM

Knowing a little about the properties of sodium helped me to understand why bicarbonate of soda is such a useful substance for domestic purposes. I even began to wish I could go back and learn a bit more chemistry. As a 14-year-old I wasn't remotely interested or motivated, probably because of the chalk and talk approach, rather than creating chemical changes and observing reactions.

Sodium exists as more than a trace element in the stars and the sun and is the sixth most abundant element on the Earth's crust, making up an estimated 2.8%. Sodium, along with potassium, is classed as a soft metal but is not found naturally as a metal. It is found in a wide distribution as compounds with other substances, the most familiar being sodium chloride or table salt. Other salts of sodium are found in many rocks and in nearly all soil types. These include halides, silicates, carbonates, sulphates and nitrates. Sodium occurs throughout the rocky crust of the Earth as feldspar, a silicate of sodium, and in various other rocks. Along with potassium, sodium is a soft metal with a silvery white lustre. It tarnishes in the air and is among the most reactive of metals.

Early civilisations knew about the benefits of sodium, although as potassium compounds were very similar in appearance they thought they were the same substance. Records exist of saltpetre from potassium being used to make glazes for pots in Mesopotamia 17th century BC and the Egyptians used sodium carbonate 16th century BC for making glass. They also used natron for embalming and preserving (see p 21).

From early times sodium carbonate was prepared by passing water through the ashes of plants which had been burned. The water was then evaporated from the solution. Soda as a term was first used for either sodium or potassium but later was used to refer to ash produced from sea plants. Potash was used for ash produced from land vegetation. The name potash, incidentally, originated from the use of large pots for evaporating the water from the solution.

Over the years soda and potash were defined as both natural and artificial products and as vegetable and mineral. Imagine playing the old favourite '20 Questions' with the ancestors – it was hard enough trying to define substances as animal, vegetable or mineral as a science game with primary school children, but at least they have a better understanding of materials through the National Curriculum. Another nugget of information that might corner your imagination for a split second: it wasn't until the beginning of the 19th century that Sir Humphry Davy decomposed both alkalis and called them sodium and potassium, which are really Latinised versions of soda and potash. The two metals cannot be isolated by normal chemical processes and were only prepared after the discovery of the electric current in 1800, when the electrolytic processes were developed. Davy's method was modified and called the Castner process and was used to prepare sodium for a long time. Most is now prepared by the Down's process which produces chlorine as well as sodium.

Sodium chemicals today are used widely in synthetic chemistry as drying agents and reducing agents. Sodium has also been used in the manufacture of photoelectric cells. Its high heat capacity and conductivity make it useful as a heat transfer medium. It is lighter than water and can be cut with a knife at room temperature. It reacts with water to give hydrogen and sodium hydroxide, so it is an extremely active chemical which easily unites with oxygen. The metal is, therefore, usually kept immersed in an inert liquid for safety. It is used in some nuclear reactors, and sodium lights give that characteristic yellow illumination to many of our roads and towns. This works by gaseous sodium glowing in a tube which has voltage passed through it.

APPLICATIONS OF BICARBONATE OF SODA

Bicarbonate of soda has many uses in a number of areas, but the most well-known domestic uses are related to cooking and household cleaning jobs. There are many other commercial applications, however, including use in some fire extinguishers, as a deodoriser and neutraliser of acids.

GOING UP . . .
Bicarbonate of soda is used in cooking as a raising agent. Bakery products are aerated by gas bubbles, either developed naturally or by being folded in from the atmosphere. The natural method of leavening bread is achieved by the addition of yeast, by bacterial fermentation or from chemical reaction, such as that produced by the addition of bicarbonate of soda. All bread produced commercially – except rye bread and salt rising/soda bread varieties – uses yeast to make it rise. Cakes, biscuits and other bakery products are leavened by carbon dioxide, which is produced when bicarbonate of soda is added. On its own, however, bicarbonate of soda makes dough alkaline and causes deterioration in flavour and discolouration, so an acid reaction substance is also added to the bicarbonate of soda to make it into baking powder.

You can find out more about this in the baking section.

ANYONE FOR LEMONADE?

Before the advent of readily available fizzy drinks, our Victorian ancestors knew how to use bicarbonate of soda to make lemonade, although you had to drink it quickly, before all the bubbles escaped. What used to be known as 'pop' (thanks to the gassy bubbles) was certainly what American children knew and loved as soda pop, originally served up through a soda fountain. Today, in France, if you order a fizzy soft drink in a café, the itemised bill may well list it simply as 'soda'. You can find some recipes in a later section.

OUT, FOUL SPOT!

Bicarbonate of soda has long been used for a variety of cleaning purposes in the bathroom and kitchen. It can be used to deodorise the fridge, nappies and rubbish bins and to clean out drains and remove soap scum. Many other uses are described later in the book.

PAIN RELIEF

Bicarbonate of soda has been used as a remedy for acid indigestion and heartburn. It makes blood and urine less acidic and increases the sodium in the body, but should not be used long term without consulting a doctor. It also relieves bee stings, nettle rash and sunburn.

SPLAT THAT GNAT!

As a deterrent to ants and fleas, bicarbonate of soda is a cheap, fairly harmless alternative to commercial products. Smelly pets can be made sweeter when cleaned with bicarb.

WHITE TEETH AND LUSTROUS LOCKS

Used in toothpaste, bicarbonate of soda helps to remove stains and freshen breath. When used during hair washing it can help remove build-up of other products and give your hair an extra shine.

STOP THAT FIRE!

Bicarbonate of soda has long been used in dry chemical fire extinguishers to put out electrical or flammable liquid fires. The bicarbonate of soda works by interrupting the flame chain reaction. This chemical agent is one of the main chemicals used as a powder in the mixtures of dry chemicals stored in pressurised tanks or containers. These can be put onto vehicles or into portable extinguishers in advance and are pressurised by an inert gas, such as carbon dioxide, nitrogen, helium or argon. When discharged they replace the oxygen in the air which would otherwise feed the fire. Carbon dioxide is the most commonly used propellant and the action starts by puncturing the seal of a cartridge of the liquefied gas. Before this method became common soda-acid fire extinguishers generated carbon dioxide by mixing a solution of bicarbonate of soda with sulphuric acid. So there we have two applications of sodium bicarbonate in dealing with fires. Portable dry chemical fire extinguishers are designed to reach 2 to 5 m (6 to 15 ft) and discharge in about 10 to 15 seconds, so the chemical reaction achieved is both impressive and effective.

NEUTRALISING ACID SPILLS

Bicarbonate of soda is commonly used for neutralising acid spills, such as battery acid.

Another surprising fact to the uninitiated, like me, is that it is used to counteract the effects of white phosphorus in incendiary bullets spreading in wounds to the body.

QUICK SOLUTION

Equally common is the use of bicarbonate of soda to increase the pH and alkalinity of water in a swimming pool or spa. It is added as a solution for restoring the balance of water where a high chlorine level exists. It can also be added to septic tanks to control pH and maintain the correct conditions for bacterial activity.

GLOSSARY

Because there are so many similar sounding names and applications for sodium compounds I thought it might be useful to describe them all in lay terms, i.e. that I understand!

BAKING POWDER

This is a mixture of bicarbonate of soda and one or more acid salts, such as tartaric acid (cream of tartar) or sodium aluminium sulphate, which creates a chemical reaction where carbon dioxide is released. A drying agent, such as cornflour, is also added. The precise mix determines whether the baking powder is a single – or double – acting powder. Baking powder has a fairly short shelf life of 6–12 months, unlike bicarbonate of soda, which keeps for longer.

Single-acting baking powder produces a chemical reaction in the bowl when the bicarbonate of soda comes into contact with a liquid. Double-acting baking powder starts the reaction in the bowl but reacts again when the mixture is heated.

BICARBONATE OF SODA

This is also known as sodium bicarbonate, sodium hydrogen carbonate, baking soda, saleratus and bread soda. It is a white, crystalline powder or granular salt which is soluble in water. It is used in the manufacture of fizzy drinks and baking powder and is also a component of fire extinguishers. Medical uses include counteracting stomach and urinary acidity, cleaning mucous membranes and dressing burns.

When used for cooking, bicarbonate of soda must be used with an acid to start a reaction. This may be provided by, amongst other ingredients, sugar, honey, yoghurt, citrus juice or fruit.

CREAM OF TARTAR

This is the most commonly known and widely used leavening acid which is contained in baking powder. If you use bicarbonate of soda as a leavener the mix needs to include an acid such as that found in sugar, yoghurt, fruit or cocoa. Cream of tartar provides the acid needed to cause a reaction with moisture in the mix, usually milk or water.

Cream of tartar is a white sediment found lining the inside of wine caskets after fermentation (tartaric acid). This sediment can be removed and used to produce a fine white powder.

PEARLASH

Potash is the ashes from roots and treestumps that have been sieved again and again with hot water and then boiled down until a residue of brown ash was left. When potash was put into an oven and continuously stirred, it would eventually become "pearlash." Pearlash was worth far more than potash and by the mid-1700s pearlash production had evolved into a major industry and exports of ash from the American colonies were established in Britain for use in the glass industry and soap factories. During the 1760s pearlash became a popular substitute for yeast in baking.

SALERATUS

The name comes from the Latin *sal aeratus* which means aerated salt, and it was used by doctors as a safe treatment for acid indigestion. In Europe chemists put carbon dioxide gas through solutions of sodium carbonate, to form less alkaline sodium bicarbonate. This became known as saleratus. Bakers in America discovered that European bicarbonate of soda was superior to the pearlash they had been using. This product was more reliable and gave less of a bitter aftertaste.

SODIUM CARBONATE
(soda ash, sodium decahydrate or washing soda)

Sodium carbonate a constituent of many mineral waters and occurs as the main salt component in sodium bicarbonate. It is used as a general cleaner and has many industrial applications. In medicine it is used to treat skin inflammation and to clean parts of the body. Sold as washing soda it is a cheap but effective household cleaning agent and useful for washing heavily soiled fabric. It is efficient at unblocking grease from drains and for cleaning sinks. It has been used in the photographic process and in brick making. Sodium carbonate should not be confused with bicarbonate of soda for cooking, as it has a pH of 11 and is therefore not suitable for consumption.

SODIUM CHLORIDE
(common salt)

This is the most abundant form of sodium. It is the main salt in seawater, making up 80% of the dissolved matter. There are extensive worldwide deposits which probably originated from the evaporation of prehistoric seas. Sodium chloride is present in all body fluids and is essential to human and animal life. Common salt is also used for preserving food and seasoning as well as in the manufacture of dyes, soaps, ceramic glazes, leather curing and the production of many chemicals.

SODIUM FLUORIDE

This is a mix of sodium carbonate and hydrofluoric acid. It is used as an insecticide and pesticide and is fatal if ingested. Curiously though, when added in minute quantities to municipal drinking water, it reduces dental decay.

SODIUM HYDROXIDE (caustic soda, also known as Lye)

This is a white, moisture-attracting solid, obtained by electrolysis from a solution of sodium chloride. It is highly corrosive to plants and animals. When dissolved in water it forms an alkaline solution, which neutralises acids in many commercial processes, including refining petroleum, soap and detergent manufacture and making paper, cellophane and rayon.

SODIUM NITRATE

This occurs naturally as Chile saltpetre and can be manufactured from sodium carbonate and nitric acid. It is used to make fertilisers, glass, pottery enamels, matches, explosives and dyes. Worryingly, to a novice like me, it is also used in meat pickling.

SODIUM PEROXIDE

This is an oxidising agent for bleaching paper, cloth, wood, ivory and for purifying the air in confined spaces such as submarines or aeroplanes.

SODIUM SULPHATE (Glauber's salts)

This was discovered by Johann Glauber in the 17th century as a natural mineral in the form of mirabilite. A neutral salt, it is used for detergents, glass and paper making and in dyeing and printing textiles. It was originally used as a laxative.

Uses of Bicarbonate of Soda Throughout History

NATRON

DEAD GORGEOUS

The Ancient Egyptians appreciated beauty so much they protected and preserved it even after death by mummification. A major substance in this process was natron, a white crystalline mixture of sodium bicarbonate, sodium carbonate, sodium chloride and sodium sulphate. It was mined from dry lake beds and from the banks of the River Nile. The area of Wadi-el-Natrun lies 23 m (75 ft) below sea level and gets its name from the natron. Known to be mildly antiseptic as well as a good exfoliant and drying substance, it was, all in all, a perfect base for adding oils and fragrances used to preserve the dead. The fragrances exemplified spirituality and beauty and served to cover up the foul odours of decaying flesh, which conjured up or summoned particular deities. So you not only had to look good but also smell divine, even in the intense heat of Egypt, without refrigeration or air-conditioning.

HOLISTIC CLEANER

The Ancient Egyptians recognised that cleanliness was of the utmost importance in the living as well as the dead. Natron was the supreme, holistic, cleaning product and was used in a variety of ways for personal hygiene and household cleaning. To ensure you were really clean and pure, you needed to smell clean and not give off bad odours that might summon up evil forces. The closest you can get today to natron is a packet or tub of commercially produced bicarbonate of soda. Many of the age-old remedies and applications of bicarbonate of soda can be traced back, like so many other seemingly modern remedies, to the knowledge possessed by the Ancient Egyptians.

MUMMY'S MULTI-PURPOSE MATE

As well as for mummification, the Egyptians were known to have used natron for:

- Ridding the home of vermin
- Cleaning the body as an exfoliant
- Cleaning teeth

- Preventing body odour
- Cleaning cuts
- Making a light to paint tombs by
- Making glass

As an ancient toothpaste, natron could be mixed with a few drops of water to create a paste. A few drops of myrrh, the resin derived from thorny desert trees in modern Ethiopia, Yemen and Somalia, could also be added. Myrrh was much prized by the Egyptians as a perfume, incense and healer, and they believed that, when it was added to the toothpaste, it served as a gum protector and improver of oral hygiene.

Added to castor oil, natron provided a smokeless fuel by which painters and artisans could work in the ancient tombs without creating harmful smoke and soot, which might spoil the work.

NATRON AND GLASS MAKING

Pliny the Elder claimed that the invention of glass took place on the Palestinian coast, when natron merchant ships were sailing from Egypt to the mouth of the River Belus near Ptolemais. The crews needed stones to hold up their cooking pots but, as these were in short supply, they crew used some of the natron as a substitute. The heat from the fire caused the natron, rich in sodium, to fuse with the surrounding sand, when the fire became too intense, and produced a new material – glass. This story has been challenged over the years, however, as the temperature of the fire would not have been great enough to cause fusion.

Another theory suggests that glass invention occurred around 2200 BC in Iran. Coloration of glass was already understood by the time Tutankhamun reigned, around 1330 BC. Coloured glass was used in furniture and architecture from then onwards, but it was the Romans who really played a major role in industrialising the process around the Mediterranean region.

ROMAN GLASS

The Romans made glass from two types of materials, one of which included natron which came from Egypt. When spring came the lakes evaporated, exposing the deposits of natron around the edges and in the lake beds. The Romans used this to produce glass on a large scale from about 500 BC to AD 850. Before this time glass was considered as a product for decorative use rather than for its many more practical uses. It was manufactured by melting alkali (potash or sodium) with silica such as quartz or sand. Different semi-precious stones gave different colours, for example turquoise gave a pale blue glass, and fluorite, a purple hue.

WORLD GLASS

During the next several hundred years, glass developed in different ways in various parts of the world, and was based on either soda or potash. The soda was found in Mediterranean regions in the ashes of plants in sea marshes and in seaweed. Germany and Bohemia used potash from beech wood and France used bracken as a source. The quality of the glass depended on the preparation of the soda.

Lead glass or flint glass was invented in the 17th century, in the days of William and Mary of Orange, in the Netherlands. This could be cut to enhance designs and show off its sheen. After this, there were still more changes and refinements to glass and its production; for example, large amounts of lead were used with potash and Venetian glass was produced with soda. Flint glass was made from three parts sand, two parts red lead and one part potash or pearl ash from Canada or Russia.

BLOWN AWAY

Europe seemed to have the monopoly on expensive and decorative blown glassware until a carpenter, in Massachusetts, USA, invented a new method of pressed glass in 1827. Then, in 1864 in Western Virginia, William Leighton brought about a revolution in the process of making glass at a fraction of the price of the lead and flint varieties. His pressed glass was of equally good clarity and could be produced thinly by being pressed into a mould. The major difference though, was the use of bicarbonate of soda and lime instead of lead. All this and manufactured at a third of the price! The bicarbonate of soda was also readily available in large quantities.

RADIOACTIVE

You may be interested to learn that one particular type of glass made in the USA in the 1880s – known as Burmese glass – contained, alongside bicarbonate of soda, some rather unusual ingredients, one of which was uranium! This was commonly used to colour yellow and green glass for over a hundred years. Understandably, it was banned during the 1940s, due to health concerns of the glass-workers and also, rather obviously, because of its use in making the atomic bomb. The British Government apparently confiscated huge quantities of materials containing uranium at the end of the Second World War from glass producers. Quantities of bicarbonate of soda were also seized.

How about this as a shopping list or recipe for making glass:

- 100 lb white sand
- 36 lb lead oxide
- 25 lb potash
- 7 lb potassium nitrate
- 5 lb bicarbonate of soda
- 6 lb fluorspar
- 5 lb feldspar
- 2 lb uranium oxide
- 1.5 pennyweights colloidal gold

. . . on reflection, perhaps not!

THE FAIREST METHOD OF ALL?

Using bicarbonate of soda for commercial production of glass has several benefits. It is effective at high temperatures of around 400°C +. During manufacture the glass gives off acid fumes and dust. Bicarbonate of soda neutralises acid components of gases so that they can be discharged into the atmosphere. The residual chemicals of sodium sulphate are recycled in the furnace, using bicarbonate of soda.

PEARLASH

As mentioned earlier, baking soda, or bicarbonate of soda, was first used by American colonists in the form of a refined type of potash called pearlash. By the mid-1700s production had evolved into a major industry and imports of ash from the colonies, with their abundant supplies of trees to burn, were established in Britain for use in the glass industry and soap factories.

Pearlash was made from potash by baking it in a kiln to remove impurities. This produced a fine white powder. During the 1760s pearlash became a popular substitute for yeast in baking – yeast took a long time to knead and was, therefore, hard work to do on a regular basis. Since pearlash was quite alkaline, it was added to the mix to counteract the sourness of sourdough breads, which were leavened from a form of yeast. As well as sweetening the dough, it speeded up the rising time by releasing carbon dioxide bubbles when heated. Pearlash created in minutes what yeast took hours to achieve. This discovery brought about big changes in the baking industry and in the home.

SALERATUS

The use of bicarbonate of soda for cooking began in the 18th century. In Europe chemists put carbon dioxide gas through solutions of sodium carbonate, to form less alkaline sodium bicarbonate. This became known as saleratus. The name came from the Latin *sal aeratus* which means aerated salt, and saleratus was used by doctors as a safe treatment for acid indigestion. Bakers in America discovered that European bicarbonate of soda was superior to the pearlash they had been using as a substitute for yeast. This product was expensive but more reliable and gave less of a bitter aftertaste than pearlash.

In America saleratus was first made in 1788 when Nathan Read of Massachusetts suspended lumps of pearlash over carbon dioxide fumes produced from fermenting molasses. By the beginning of the 19th century brewers had started producing saleratus as a by-product, using the carbon dioxide released during fermentation. As the imports from Europe were expensive, the challenge was to provide an everyday, cheap product in the USA. There was also the problem of a slightly hit and miss approach, as the bubbles were often released before the dough made it to the oven. Someone needed to invent a more stable way of adding bicarbonate of soda for baking.

DEVELOPMENTS IN LEAVENING AND BAKING

BAKING POWDER

The first baking powder compounds were created with cream of tartar and bicarbonate of soda. The mixture gave more consistent results than plain old bicarb, but didn't keep as long and was more costly to use. Experiments continued with calcium phosphate, which still released most of the gas before the cooking stage. Later, sodium aluminium sulphate was discovered and, since it only reacted with the bicarb when heated, bicarbonate of soda, cornflour and the new sulphate were used to much greater effect.

In 1843 a British chemist called Alfred Bird made another version of baking powder for his wife Elizabeth, who couldn't eat yeast or eggs. This was so successful that he received an order to supply his fermenting powder to the army. This meant that soldiers could have fresh bread, and the sick and wounded could eat bread, light cakes, light puddings and other articles of food suited to their condition. He then founded Bird and Son Ltd, who, presumably, went on to make a well-known custard powder.

Later refinements resulted in the production of single – and double-acting powders. These meant that there was a lot more control over using baking powder in cooking. You can learn more about these in the chemistry and baking sections.

MINERAL AND FIZZY DRINKS

HUBBLE, BUBBLE . . .

By the 17th century people were attempting to make refreshing drinks and artificial mineral waters. At first these were required to enhance the experience of medicinal bathing in spas and springs. You could only drink mineral waters if you lived close enough to travel to the source, so there was an opportunity for clever entrepreneurs to invent a substitute and corner potentially huge markets. By the 18th century companies were producing artificial mineral water in France, Germany, Britain and Switzerland. In 1788 the 'Geneva apparatus' was used for the first industrial production of artificial mineral water. This was achieved by obtaining carbon dioxide from a chemical reaction between sulphur and bicarbonate of soda, which was then added to a mixing vessel with added salts. But the carbonisation did not take place under pressurised conditions, so only a small amount of gas stayed in the water.

In Britain, around 1770, a chemist called Priestly also discovered how carbon dioxide could be added to water. He lived near a brewery, which led him to an understanding of the nature of effervescence found in mineral water. Carbonated drinks today owe much to his early experiments. By 1821 a German doctor and pharmacist in Dresden called Struve had succeeded in inventing a much more sophisticated method of producing artificial mineral water, which resembled the make-up of waters from the Niederselters springs. These became known as seltzers and soon afterwards the production of cheaper soda water was established. If this rings bells it may be because Alka-Seltzer, a popular remedy for headaches, indigestion and hangovers in the 20th century, was made from aspirin, bicarbonate of soda and citric acid. The 'Alka' came from alkaline.

LEMONADE

What we understand nowadays as lemonade, a very sweet, sticky, fizzy tooth rotter, much loved by millions, started life as a flat mineral water flavoured with lemons and sweetened with sugar. Mass production of carbonated drinks started during the mid -19th century and the term 'lemonade' was used to describe all fruit and berry drinks. You can see some early recipes for these in the section on food. The Victorians were very fond of home-made lemonade, as described by Mrs Beeton and others. The following is based on a 19th-century American recipe for lemonade:

LEMONADE
Boil two pounds of white sugar with one pint of lemon juice. Bottle and cork. When required, put one tablespoon of the syrup into a tumbler three-quarters full of water. Add one-third teaspoon of bicarbonate of soda and drink quickly.

The Victorians also made up powders, which could be conveniently carried on a picnic or when travelling, and reconstituted in a glass of water as required:

GINGER BEER
Put into blue papers 30 grains of bicarbonate of soda and 5 grains of powdered ginger with a drachm of powdered sugar. Put into white papers, 25 grains of cream of tartar. Put one paper of each kind to half a pint of water.

(A drachm corresponds roughly to one teaspoon, and 30 grains to half a teaspoon, so 5 grains would be about a twelfth of a teaspoon.)

SOAP

EARLY BATH?
The most common early soaps were made from potash and pearlash. Early references to the use of soap include the Babylonians, about 2800 BC, and the Phoenicians, c.600 BC. The Egyptians used natron, as described above, and Spaniards and other Mediterranean people were using burned seaweed to provide the alkali they needed. Early uses included the cleaning of wool and cotton fibre prior to spinning and weaving.

LEND ME YOUR SOAP!
Pliny described Roman soap as being made from goat fat and wood ash, with salt being added to harden it into bars. Excavations in Pompeii have revealed a soap factory with bars of soap. This would have been for textiles rather than personal hygiene, as the Romans favoured olive oil and sand to clean their bodies after a good steam.

THE VERY FIRST SOAP?
Soap is basically a reaction between fatty acids and an alkali. When fats or oils are mixed with a strong alkali the fats are split into fatty acids and glycerine. The sodium or potassium in the alkali joins with the fatty acid as the basis of soap.

The three main stages of early soap making were:

- making a wood ash solution called lye,
- cleaning the fats,
- boiling the fats with the lye to make soap.

Early manufacturers obtained the lye from putting potash into a bottomless barrel over a stone slab, resting on rocks. Straw or sticks were put in the bottom as a sort of sieve and underneath that was placed a collecting container. By slowly pouring water over the ashes you could produce the lye, a brown liquid dripping down into a container below. This could then be used with the rendered down fat to produce soap. When wood supplies dwindled towards the end of the 18th century, pearlash manufacturing started to decline, making way for more commercial methods. The Leblanc process, described below, changed soap making for ever. Sodium alkalis made harder, better soap without the necessity to add salt.

COMMERCIAL PRODUCTION OF BICARBONATE OF SODA

THE LEBLANC PROCESS
As greater supplies of soda were required for glass-making and soap manufacturing, as well as baking needs, the supplies of ash declined. By the 1700s production was

uneconomical and Europe was being deforested. Potash had to be imported from North America, Russia and Scandinavia, and Louis XVI in France offered a prize for a method of producing alkali from sea salt. In 1791 Leblanc found a solution and by 1800 soda ash was being produced at the rate of 10,000 to 15,000 tons per year.

HOW BROWN WAS MY VALLEY?

After a repeal of a tax on salt in Britain in 1824 the development was unstoppable and by 1870 the output from Britain had reached 200,000 tons a year: more than all other countries put together. The price was environmental devastation. Sulphuric acid released hydrochloric acid gas into the atmosphere. For every eight tons of soda ash produced, seven tons of calcium sulphide waste hung around on the ground, smelling of rotten eggs. The surrounding land was scorched and fields and gardens produced little that was not spoiled. The people who worked in these factories or who lived close by must have suffered greatly. Luckily another process was invented without such major disadvantages.

THE SOLVAY PROCESS

Belgian Ernest Solvay invented a process whereby the only waste product generated was calcium chloride. By 1900, 90% of world production used this method. Solvay chemical plants produced about three-quarters of the worldwide requirements, which in 2005 were estimated at 41.9 billion kilograms.

TRONA DEPOSITS

However, in 1938 in Wyoming, USA, large natural deposits of a mineral called trona were discovered – and sodium carbonate can be extracted more cheaply from trona than by the Solvay method. The name is, as you may have noticed, almost an anagram of natron. I think we've come full circle!

Cleaning Solutions with Bicarbonate of Soda

HOUSEHOLD CLEANING

Bicarbonate of soda is a cheap, natural cleaning product. It won't break the bank to buy, but it will do the job. Even more importantly, it won't cause harm to the environment in the same way that a number of expensive, caustic products do. Being a natural sceptic, I didn't believe how effective some of the following suggestions could be until I tried them. I'm all in favour of using as few chemicals as possible, as a matter of principle, and the extra effort involved in preparing the pastes and solutions was minimal. As you might also have gathered, housework is not top of my agenda and for some of the kitchen and bathroom jobs I need the incentive of quick, efficient methods! Bicarbonate of soda helps to dissolve grease and dirt and also prevents smells from hanging around.

If you want to make up a solution as an all-purpose cleaner, take two teaspoonfuls of bicarbonate of soda to 570 ml (1 pt) of warm water. Adding 125 ml (4 fl oz) of white vinegar will make an even better cleaner, but beware of the gaseous effect of adding bicarbonate of soda to vinegar and allow room for expansion. After the fizz has died down you can pour the mixture into a screw-topped bottle, ready for use.

CAUTION
As with all products and cleaning solutions, you should try out the suggested method on a small area first. I make the following suggestions in good faith and I can't be held responsible for any adverse effects.

First and foremost, don't use bicarbonate of soda on non-stick pans, or you're likely to ruin the surface.

NOTE: In the following, whenever vinegar is mentioned please use white vinegar unless told otherwise.

KITCHEN USES

WORKTOPS

Use bicarbonate of soda on a damp cloth to clean worktops and leave them free of smells. You can sprinkle the powder directly onto a cloth or sponge, or make up a solution. Rinse with clean water and dry. The solution can be used to clean your cleaning cloths as well and to stop them smelling sour.

STUBBORN STAINS ON WORKTOPS

To remove stubborn stains from marble, plastic or formica, make a strong paste of bicarb with water. Scour gently and then rinse thoroughly.

CLEANING THE EXTERIOR OF WHITE GOODS

All exteriors of white goods, such as fridges, washing machines and freezers, will benefit from a wipeover now and again with a cloth soaked in a solution of bicarbonate of soda and warm water. Greasy marks and dust will quickly disappear.

DISHWASHERS

Bicarbonate of soda will remove that nasty, crusty build-up you get on the inside of the door. Use as a paste and rinse off before running the machine. Try using bicarbonate of soda instead of the usual powder or tablet once in a while to clean the dishwasher, too.

DISHWASHER POWDER

Instead of using a commercial brand of dishwashing detergent, try mixing two tablespoons of bicarbonate of soda with two tablespoons of borax for a load.

MICROWAVES

Use in solution to clean up food splatters on the base, sides and ceiling. Remove the turntable plate and wipe over the interior, taking care not to let any water drip through the holes in the walls. The turntable can be cleaned with the same solution.

CLEANING THE FRIDGE

I've always used bicarbonate of soda in solution to clean the inside of the fridge. With fridges you had to turn off and defrost from time to time, I always wiped the trays and shelves with a strong solution. As well as cleaning up greasy marks it removes any odours from spills, milk bottle bases and cheesy whiffs. With the larder fridge we have now, I remove any uncovered food, wipe down the shelves and leave to dry as far as possible before replacing food items. Alternatively, wipe down with paper kitchen towel before putting the food back in.

If you are planning on turning off your fridge while you are away on a long holiday, or are moving house and need to leave it switched off for a while, using bicarbonate of soda is a good way to save time and prevent mould forming. I use it to clean out the fridge-freezer in our home in France when we leave each time, leaving the doors open and the power off. The only time we went back to black mouldy deposits in the freezer was when a helpful workman closed the freezer door (to be fair, he did think he was doing us a favour) and I hadn't given the shelves the treatment. On return a quick wipe over with a solution of bicarb in cold water was all that was needed before starting up the unit again.

Any stubborn stains that have sat at the back of the fridge for a while may be removed with a tablespoon of bicarbonate of soda mixed with a squeeze of toothpaste.

FREEZERS

Another chance to use the solution is when the spare freezer in the garage has been off, or we've stored various family members' freezers etc. Cleaning with bicarbonate of soda and warm water has always kept the units fresh – as long as the door is kept propped open.

If you defrost the freezer, wait for all the ice to melt then wipe carefully with warm water solution and allow it to dry thoroughly. Restart the freezer . . . and close the door!

CLEANING THE DRAIN

If your kitchen drain tends to block because of a build-up of grease or soap, try pouring about a quarter of a tub of bicarbonate of soda down it every month as a preventative measure.

BLOCKED DRAIN?

If the tip on the previous page hasn't worked, or you haven't bothered to check for a while, use half a tub of bicarbonate of soda with a cup of warm vinegar. Leave this to work for an hour or two, then pour down lots of very hot water.
An alternative is to use equal measures of bicarbonate of soda and table salt, washed down with boiling water.

CLEANING THE BIN

Use a fairly strong solution of bicarb and warm water to wash the kitchen bin out. It will keep the smell down as well. This is good for all plastic and stainless steel bins.

REMOVING PLASTIC WRAPPERS FROM HOT SURFACES

Have you ever caught a corner of a plastic bag wrapper of a loaf of bread on the toaster? Can't say I have, but it must be hard to get it off again. Dampen a cloth and make a mild scouring pad with some bicarbonate of soda. Apparently, it will work well. Obviously you need to unplug electrical goods such as toasters before cleaning them like this and avoid getting any moisture on the inside element.

REMOVING NEWSPRINT STAINS FROM SURFACES

Now this is one that I have experience of. The print can be a real pain to get rid of if you leave a newspaper lying on a slightly damp surface, or a drop of rain gets the paper wet before delivery. Make a paste with bicarbonate of soda and a little water and rub gently until the stain disappears.

STAINLESS STEEL SINKS

Sometimes stainless steel can appear to be a bit of a misnomer, especially when tea and coffee dregs go down the sink. I regularly clean mine with a solution of vinegar and bicarb, as above.

An alternative to using vinegar with bicarb is to use soda water, which will avoid the possibility of scratching, especially on new sinks and appliances. If in doubt though, just use warm water.

REMOVING HARD WATER STAINS

Water stains are another common occurrence, if you live in a hard water area. Use bicarbonate of soda, sprinkled onto the stain, and then wipe clean with vinegar. Watch out for the fizz, remember!

STEEL WOOL PAN SCRUBBERS

Despite the many other scourers available these days, some of us still like to use the old pads on stubborn stains to baking pans and cast iron. If you don't use them very often, keep them in a box with some dry bicarb powder and they won't go rusty.

CLEANING THE BREAD CROCK OR BREAD BIN

We have a splendid, French deep bread crock which weighs a ton but keeps bread fresh, as long as the odd piece of organic bread isn't left lurking in the bottom out of sight: in which case the jolly green mould is very smelly, especially if left for a few days in a nice warm kitchen. Washing out with a solution of bicarbonate of soda in warm water keeps it mould free and sweet smelling. Leave to dry thoroughly and cool down before storing bread again.

CLEANING OUT A FOOD PROCESSOR WITH DRIED-ON FOOD

Of course, this will never happen to you, but some of us are natural slobs. There are always times when the bowl is left in the sink, without soaking, and the blade is stuck away behind a mountain of pots too big for the dishwasher. If you haven't used your blender for a while it might need just freshening up. Use a couple of teaspoons of bicarb with a cup of warm water, put the lid on and run it for a few seconds. Afterwards you can wash as normal, or simply rinse and dry.

REFRESHING A VACUUM FLASK

To clean out an old vacuum flask add a teaspoon of bicarbonate of soda and fill with warm water. Leave to steep for a while, then rinse out with cold water.

POTS AND PANS WITH BAKED-ON DEPOSITS (NB: not non-stick!)

1. For really burnt-on grease and grime you could try the full assault. Remove what burnt-on food you can, then pour a layer of bicarbonate of soda onto the base of the pan and use just enough water to moisten the soda. Leave until the next day, then scrub clean. Whenever I make custard the milk always burns on the bottom of the pan. This method makes cleaning a lot easier.

2. If that doesn't get it all off, for example after a pan has been baking away in the oven for hours, put a layer of water in the bottom and add a cup of vinegar. Bring this to the boil, remove from the heat and then sprinkle in two tablespoons of bicarbonate of soda. Remember to expect the fizzing effect, and stand back unless you fancy the idea of two nostrils' full of the smell of vinegar at close quarters. The remains of the cooking should then loosen off without further resistance.

3. If the pan isn't too bad, soak it in bicarb and water for half an hour before washing normally. Alternatively, if you're not bothered about scratching the pan, use the soda dry with a scouring pad.

4. Here's another idea: mix one spoonful of salt with two of bicarbonate of soda and mix with a little lemon juice. Spread this over the offending article and leave to dry. Rub off with a scouring pad.

REMOVING STAINS ON ENAMELLED CAST-IRON WARE
Most of my casseroles and saucepans are of the cast-iron variety; perfect for using on, and in, an Aga (making for minimal effort – the same pot for cooking on top, inside the oven or serving from at the table). The most loyal of these, given to me by my mother after many years of service with her, is now about 30 years old. The inside of the pot had become markedly browner than the original creamy colour over the years, as a result of daily use and food stains (and regular burnt-on food in the early days of

Aga cooking). It has been given a new lease of life by a treatment of bicarb and vinegar. Use Method 1, as opposite. Some of the newer pots won't get this treatment though, because they have a different, dark, non-stick coating.

GAS BURNERS

If you can remove the gas burners easily to clean around them, immerse them in a pan of boiling water with four tablespoons of bicarbonate of soda for a few minutes.

CLEANING THE TOP OF THE STOVE

Wet the area and sprinkle bicarbonate of soda onto the top. Allow to sit for about 30 minutes before removing and rinsing off.

CLEANING THE OVEN INTERIOR

Even if your oven is really dirty you don't need to go to the desperate lengths of buying expensive and toxic oven cleaners. All you need is a thick paste of bicarbonate of soda and water. Wipe over the inside when the oven has cooled down. Leave overnight while the soda does its work. Wipe down with warm water to remove the grease and grot. Stubborn areas can be treated with neat bicarbonate of soda on a scourer or cloth. You can clean a glass door with a weak solution, perhaps kept in a labelled spray bottle.

An alternative method is to spray the inside of the oven (once it is cool/cold) with water and place bicarbonate of soda on the oven floor. Keep the oven moist by spraying every few hours and then leave overnight before removing the deposits.

GLASS AND STAINLESS STEEL COFFEE POTS (not aluminium)

Fill the pot or jug with warm water while it is standing in a sink. Add a dash of white vinegar and half a teaspoon of bicarbonate of soda. Leave to fizz and soak for a few minutes. Rinse with fresh water and dry carefully.

COFFEE MAKERS

Running bicarbonate of soda and water through the machine will give it a good clean and refresh it. Rinse with plenty of clear water afterwards, to remove all traces of the bicarbonate of soda.

BABY BOTTLES

These can be cleaned with a solution of bicarbonate of soda and water. Again, be sure to rinse thoroughly.

REMOVING STAINS FROM COFFEE MUGS

Sprinkle a little bicarbonate of soda into the bottom of the cup and add warm water to make a paste. Leave to stand for a while before rinsing.

STAINLESS STEEL AND CHROME KITCHEN APPLIANCES

These can be polished with a moist cloth and bicarb.
I tried it on our kettle which stands on the hob, but make sure the kettle is cool or empty first, as otherwise the bicarb dries too quickly for you to polish the marks away.

REMOVING GREASE SPILLS

If you spill grease or oil on the kitchen floor while cooking, mop up what you can with paper towels, then sprinkle bicarbonate of soda onto the spill. This can be mopped up later when you have more time, and will prevent the 'skidding' that occurs if you use detergent straight away and which spreads across the floor.

WASTE DISPOSAL UNITS

Do houses still have these? If so, you can apparently clean them out with a mixture of bicarbonate of soda, lemon juice and vinegar. I prefer composting my kitchen waste.

BATHROOM USES

Anywhere that water lurks there are likely to be odours and mouldy deposits. Add a lack of air circulating, as for example in a bathroom, cloakroom, toilet or shower room, and the possibilities for using bicarbonate of soda are endless, it seems.

SHOWER CURTAINS

I don't have these anymore, but remember vividly that smell of old plastic mac and camping toilets, with a trim of mould along the seams. Put them in the bath and brush or sponge with four or five tablespoons of bicarbonate of soda. You can use an old toothbrush for the seams. Rinse with hot water and white vinegar. Hang the curtains – outside if possible – to dry thoroughly.

CLEANING TILES AND GROUTING

This gives instant results and a sense of extreme satisfaction. Using an old toothbrush and a paste, made of two parts bicarbonate of soda to one part white vinegar, rids the grout of that yellowy-pink look that builds up in the shower bit by bit. The initial smell of vinegar quickly disappears, so you won't smell like a chip shop. Rinse with the shower head for quick results.

CLEANING TAPS

Another gratifying job! While you have some of the paste mixed for the tiles, as above, give the base of the taps a treat as well. I'd been trying to get the marks off where the basin meets the chrome for a while and had even bought a commercial product, which I didn't use when I saw the contents and read the cautions attached. The bicarb worked a treat.

CLEANING OUT THE BATH AND WASH BASIN

Using the paste of two parts bicarb to one part vinegar, you can remove the built-up soap scum by gently applying with a damp cloth. Rinse with cold water. The plug hole will benefit as well.

FRAGRANT SINK CLEANER

Put about four tablespoons of bicarbonate of soda into a bowl with enough liquid soap to make a paste. Add a drop of tea tree oil, two drops of lemon oil and two of orange or lavender oil.

CLEANING BATHROOM TRAPS

If you have one of those removable traps in the shower, bath or wash basin, you can remove it, along with the gross, slimy build-up of hair, shampoo, conditioner, gel, etc. that is so foul to deal with. Soak the removable bit in a plastic tub in a solution of bicarb and warm water. Add vinegar for extra effect (use an old ice cream tub, but make sure you throw it away or recycle it immediately afterwards).

GETTING RID OF DAMP UNDER THE BASIN

Where cleaning cloths lurk, along with unwanted or half-used bottles of something you got for Christmas last year and the odd flannel or old toothbrush, is the perfect place for nasty damp patches to form. Leave a tub of bicarbonate of soda with the lid off to absorb the damp.

CLEANING THE TOILET PAN

Instead of using bleach, try putting bicarbonate of soda down the pan. Leave a while before flushing away. The toilet seat can be wiped with bicarb in solution as well.

FLUSHING WITH SODA

Every month or so, sprinkle a little bicarbonate of soda into the cistern and leave overnight. Flush in the morning and you, too, will flush with pride!

REMOVING SOAP SCUM FROM THE SHOWER

If you have a shower cubicle you will know how difficult it can be to get in all the corners and to clean the inside of the sliding doors properly. Try using a paste of bicarbonate of soda and vinegar. Use four tablespoons of bicarb to two of vinegar. This can be applied to the doors with a cloth and you can use an old toothbrush to get into the corners. It will cut through the soap scum and refresh the darkest recesses. It also refreshes the plug hole and drain, as a bonus.

GENERAL USES

Remember, it's always a good idea to carry out a spot test on a surface if you haven't used the suggested methods before. If you have left-over solutions, make sure they are labelled and kept out of harm's way.

FLOOR CLEANER

Use a solution of bicarbonate of soda dissolved in warm water. Use as you would any other floor cleaner.

CLEANING UP VOMIT

My mum told me about this when our children were babies. It works really well on most floor surfaces. Use a solution of bicarbonate of soda and water for mild cases and work into the carpet gradually. It kills the lingering smell too.

SILVER CLEANER 1

Silver can be cleaned with a paste of bicarbonate of soda and water, which saves you using toxic chemicals again and also prevents your precious silver being worn away. Use three tablespoonfuls of bicarb to one of warm water. Spread over the object and let it dry. You can use an old toothbrush to get into all the crevices. Polish with a soft, clean cloth.

SILVER CLEANER 2

For heavily tarnished silver, try this method. The tarnish is silver sulphide and this can be restored without abrasive cleaners through a little alchemy. For large items fill a plastic bucket with hot water and put a piece of aluminium foil in the bottom. Sprinkle bicarbonate of soda over the item to be cleaned and put it into the water. Leave it for 15 minutes before taking it out and drying with a soft cloth. Smaller items can be washed in an old aluminium foil baking tray in the same way. Magic!

(Well, okay, not really magic, but a reaction between aluminium, a very active metal, and silver which is inert. Silver reacts with sulphur in the air, which makes it tarnish. The aluminium removes the sulphur by chemical reaction.)

BRASS CLEANER 1

A recipeI came across for cleaning brass suggests using two tablespoons of flour, one tablespoon of salt and one tablespoon of bicarbonate of soda, mixed together with vinegar to make a thick paste.

BRASS CLEANER 2

Sprinkle some bicarbonate of soda onto half a lemon and rub clean. Rinse and polish dry thoroughly.

CRAYON MARKS

These can be removed with a mild solution of bicarbonate of soda and water, rinsed with hot water.

INK SPILLS

These can be treated on a hard floor surface with bicarbonate of soda and vinegar solution. Rub gently at the stain until it fades.

CARPET CLEANER

Whatever the spill, use bicarbonate of soda to soak up as much of the liquid as possible. This can then be swept up and the carpet dabbed with a mild solution of bicarbonate of soda and warm water. Allow to dry thoroughly before vacuuming.

HELPING TO PUT ON RUBBER GLOVES

This sounds like a good trick, only I always forget to put the gloves on before I start a grotty job. After you have used the gloves, rinsing them before removal, sprinkle a little bicarbonate of soda inside each one. It will help to get them on next time and will prevent any smell.

ALUMINIUM WINDOW AND DOOR FRAMES

Mix two parts bicarbonate of soda to one part vinegar for cleaning tracks, fittings and frames. Wipe clean with a damp cloth.

VINYL CHAIRS

Simply wipe down with a weak solution of bicarbonate of soda and water. Follow this up with a dry cloth.

SPILLS ON BOOKS

Horror of horrors, a spilt drink has spoiled your favourite book! Sprinkle bicarbonate of soda on the wet pages and let them dry – in the sun if possible. Of course, if it's not your book, you'll have to own up (or, at least, come clean!).

SUPER FILLER

Apparently, you can mix bicarbonate of soda with superglue to make a filler for cracks. Since I won't use the latter because of the smell and chemicals involved, I can't say I have any experience of this technique. Careful with that glue, though!

LAUNDRY USES

If in doubt about the colour fastness of the item, or in the suitability of using bicarbonate of soda on delicate fabrics, always try a small sample first on a less obvious part of the garment or item.

FABRIC SOFTENER

Added to the laundry, bicarbonate of soda acts as a fabric softener. Mixing one part vinegar (white), one part bicarbonate of soda and two parts water will make a suitable solution which you can keep in a bottle (labelled). Add four tablespoons to the final rinse.

SILK UNDERWEAR

This has been recommended for use with silk underwear, when you don't want to use commercial fabric conditioners that may harm sensitive skin. Rinse the garments in a mild solution of bicarbonate of soda and water every ten or so washes. This should remove any build-up of minerals without affecting the skin.

CLEANING THE IRON

Deposits on the base of an iron can be removed with a paste of bicarb and water. Unplug the iron first and let it cool down!

STAIN REMOVER

Rinse the article in cold water before the stain dries. Apply bicarbonate of soda to the stain and then wash as normal. Don't try this on delicate fabrics such as silk (but see above for silk underwear treatment).

WHITER NAPPIES WITH LESS EFFORT

If terry nappies make the comeback they deserve, to save using valuable resources, then this will help. Soak the nappies in a bucket of bicarbonate of soda solution overnight. When you come to wash them, you won't need so much washing powder or much hot water to keep them white.

WHITENER FOR WASHING

Mix one part lemon juice with one part bicarbonate of soda. Add to your whites for a mild bleaching effect.

REMOVING BLOODSTAINS

Soak clothing in a bucket of cold water with a cup of vinegar and an equal amount of bicarbonate of soda. Leave overnight before washing as normal.

REMOVING VOMIT

Soak as for removing bloodstains, but with double the bicarbonate of soda, and leave out the vinegar.

GREASE STAINS

These can be removed, before you wash the item, with a paste of two parts bicarbonate of soda to one part cream of tartar, mixed with a little water. Rub the grease mark gently before washing.

REMOVING MARKS ON SUEDE AND UNWASHABLE FABRIC ITEMS

Sprinkle bicarbonate of soda over the item and brush gently to remove dirt and grease.Check a small part of the item that won't be obvious first.

OUTDOOR USES

DEALING WITH GREASE OR OIL SPILLS

If you have grease or oil spilt on the garage floor, sprinkle with bicarbonate of soda and leave it to stand. Brush up and rinse.

OUTSIDE DRAINS

1. Outside drains need some care as well. Use half a tub of bicarbonate of soda and a cup of vinegar. Stand back as the fizz works its magic. Leave this for an hour or two, then pour down lots of very hot water.
2. You could also use equal quantities of salt and bicarbonate of soda with some boiling water.

DUSTBINS

Sprinkle half a tub of bicarbonate of soda into the bottom of the empty dustbin. Swill round with a couple of bowls of warm water and leave for a while. Empty out down the nearest drain.

PADDLING POOLS

Children's paddling pools can be cleaned with a solution of bicarbonate of soda and water. At the end of the season repeat this method and allow the paddling pool to dry thoroughly before putting away in storage.

SPA BATHS AND POOL AREAS

Use a solution of bicarbonate of soda and water to wash down the walkways.

WINDSCREENS AND WINDOWS

These can be cleaned with a damp cloth sprinkled with a little bicarbonate of soda. I still like vinegar for the job, myself.

BARBECUE GRILLS

These get a lot of hammer. Sometimes the grease and dirt can be burnt off by the next barbecue, especially if these are a frequent summer occurrence. 'In our climate?', you shout. More often than not, rain stops play in our garden (we – like you, probably – are famous for wet barbies). A scrub with bicarbonate of soda and vinegar will clean up the mess, even after the rain.

FIRE PREVENTION

As bicarbonate of soda creates carbon dioxide, it is a useful substance to have around where there are flammable chemicals or grease. The soda will prevent oxygen feeding the flames. Keep a tub handy.

GARDEN FURNITURE

Wipe down with a solution of water and bicarbonate of soda at the end of the winter to clean and freshen the table and chairs.

GREENHOUSES

Our old greenhouse gets quite musty if there isn't enough ventilation. Ever since water got through the roof we've had a hard time getting the wood dried out and the air freshened in the winter months. Half a tub of bicarbonate of soda left open absorbs some of the damp smell and creates carbon dioxide for the plants. You can also wash down the interior with a solution of bicarbonate of soda and water to inhibit spores and bacteria.

DOORMATS

Porch and outside doormats can be cleaned with a sprinkling of bicarbonate of soda. Brush vigorously and then sweep away the dirt. Next time it rains the job will be complete.

MELTING SNOW AND ICE

Instead of using salt on the garden path, which will harm any plants it comes into contact with, use bicarbonate of soda to melt snow and ice.

REGULATING THE pH IN A POOL OR SPA

When alkalinity needs to be increased in swimming pools and spas, bicarbonate of soda can be added to restore a balance if there is too much chlorine.

SEPTIC TANKS

Adding a tub of bicarbonate of soda to a septic tank will maintain a healthy pH and keep an environment just right for bacteria to do their job. I'm still glad that we are on mains drainage now, though!

BATTERY ACID SPILL AND CLEANER

Sprinkle bicarbonate of soda onto battery acid spills to neutralise the acid. You can clean battery terminals with two teaspoons of bicarbonate of soda mixed with a litre (2 pt) of water. Smearing a little Vaseline around the base of the terminal will help prevent further problems.

TEST FOR SOIL ACIDITY

Use bicarbonate of soda in the garden to test for acidity. If the soil bubbles it is too acidic and will need neutralising.

WASHING BUGS OFF THE CAR

Use a solution to wash the car and remove dead insects. Use on the windscreen to remove grease and build-up of dirt.

SPRAYING AGAINST POWDERY MILDEW

This is an old treatment to prevent plants suffering from milldew. It is also a lot safer to use than harsh chemicals, especially if you are trying to be organic. Before spraying, first water the foliage and let it dry. Spray in the evening and never in full sunlight, otherwise the leaves will probably get scorched.

Mix together half a tablespoon of bicarbonate of soda with half a litre or a pint of soapy water in a spray bottle. Apply to both sides of the leaves of plants normally affected.

BLACK SPOT PREVENTATIVE

Use the spray above, but add a drop of vegetable oil to the bottle to help the mixture stick to the leaves when dried. The soap helps to spread the mixture and the bicarbonate of soda makes the surfaces of the leaves alkaline, which will inhibit the fungal spores. The biggest advantage of this is that there will be no adverse environmental impact because the ingredients are safe.

PAINTBRUSH RESTORER

You may be able to salvage your old paintbrushes that have gone hard by soaking them in a pint of hot water with two tablespoons of vinegar and four of bicarbonate of soda. This won't work on brushes used for gloss paint, which you didn't clean with white spirit first, but will renovate emulsion brushes.

PETS AND PESTS

RIDDING YOUR HOUSE OF ANTS

Put some bicarbonate of soda inside the bottom of the door frame to drive away ants. Once they get in the house they are a real pain.

DOG BATH AND FLEA CONTROL

Adding a little white vinegar and bicarbonate of soda to a bath will make your dog less doggy and also leave a soft coat. Some suggestions I've come across advise sprinkling bicarbonate of soda directly onto the dog's coat, but I'm not a dog owner and it sounds like a bit of a trial to me: for the dog and the owner. If left in contact for long the soda might harm the dog's skin, so the bath seems a kinder and simpler way.

Sprinkling into the dog's coat and combing out sounds like a dry shampoo treatment.

CLEANING OUT A FISH TANK
Having removed fish, water and other contents, use a thick paste of bicarbonate of soda and white vinegar to remove stains and smells. Rinse thoroughly before reintroducing the fish.

CLEARING UP SOMETHING THE CAT BROUGHT UP
Using bicarbonate of soda in water is a great way to wipe over surfaces where the cat has parked his lunch or left his calling card. As with carpet cleaning and other floor surfaces, it removes the stain and takes the smell away, too.

COCKROACHES
Luckily I've never had to deal with these, but apparently mixing equal parts of bicarbonate of soda with sugar and spreading where they lurk, i.e. dark, dank undersink areas and behind skirting boards or kicking boards, will see them off.

NEUTRALISING NASTY NIFFS

SWEETER SMELLING RUBBER GLOVES
After you have used the gloves, rinsing them before removal, sprinkle a little bicarbonate of soda inside each one. It will help to prevent any smell coming from them when you use them next time.

PILLOWS
If your pillows are liable to smell and you can't give them a good airing outside, put a little dry bicarbonate of soda inside with the filling to keep them smelling fresh.

MUSTY MATTRESSES
If the house, or spare room, has been shut up for a while and not properly aired, the mattress may smell a bit damp and musty, even if it is perfectly dry. Turn the mattress and sprinkle a little bicarbonate of soda on before you make up the bed with fresh bedding.

AIR FRESHENER FOR CARPETS
Equal proportions of bicarbonate of soda and cornflour can be sprinkled on a smelly carpet or rug at night. In the morning, vacuum up the powder and the whiff should be gone.

AIR FRESHENER FOR THE ROOM
Mix a few drops of preferred oil fragrance with a small bowl of bicarbonate of soda. This is much cheaper than a commercial air freshener, without any cloying or

overpowering fragrance. You could even use an ashtray, thereby indicating that smoking is not encouraged in the house!

GETTING RID OF A WOODY SMELL

If you don't like the smell that emanates from that old wooden chest or wardrobe, try sprinkling a little bicarbonate of soda in the bottom and covering it with lining paper. Alternatively, place an opened box or tub of the powder in the bottom. Don't forget it is there, though – you might end up knocking it over!

SMELLY SWIMMING BAGS

These can be freshened with a sprinkling of bicarbonate of soda.

DEODORISING THE BIN

Sprinkle bicarbonate of soda in your dustbin to keep the smell down in hot weather.

UNDERSINK ODOUR

Place an open container of bicarbonate of soda on the shelf under the sink to avoid nasty niffs from cloths and cleaning paraphernalia. Of course, if you rinse out the cloths in bicarb, they won't smell anyway.

ATTICS AND CELLARS

Again, open containers of bicarbonate of soda will act as air fresheners in areas you don't visit very often, like the loft.

FRIDGE DEODORISER

You can do this by either wiping out the fridge with a cloth washed in bicarbonate of soda solution, or by leaving a small opened pot of the powder inside.

DRAINS AGAIN

Yes, as well as cutting through the grease, the bicarbonate of soda will deodorise your drains.

GREENHOUSE MUSTINESS

The use of an open, half tub of bicarbonate of soda will freshen the air during the months when you can't have the windows open.

VACUUM CLEANER TREAT

Sprinkling a little powder into the new bag will stop that old dog smell, especially if you don't need to change the bag very often.

'HANG ANYWHERE' DEODORISING SACHETS

This recycling trick might be useful and mean that you don't have bowls of white powder lying around everywhere. Use the foot of an old stocking or pair of tights to make a deodorising sachet that you can hang anywhere. Put the bicarbonate of soda into the foot, twist the leg round a few times and go back over the ball of powder with a second layer of tight or stocking. This can be tied with a string, ribbon, elastic band, or whatever grabs you, and hung in a cupboard, hidden in the dog basket, or attic.

SMELLY SHOES

The worst smell in the world comes wafting out of my son's five-a-side trainers. Sprinkling bicarbonate of soda into them – if he leaves them here again – will be an undiluted pleasure!

DEODORISING WOODEN CHOPPING BOARDS

The smell of onions etc. can be eradicated by wiping the board with a solution of bicarbonate of soda and warm water.

NO-SMELL KITTY LITTER

Sprinkling a little bicarbonate of soda into the litter tray, when washing out or changing the contents, will deodorise the tray.

DEODORISING DISUSED FLASKS

As well as cleaning, a solution of bicarbonate of soda and water will refresh the inside of the flask and eliminate any residual smell.

GETTING RID OF ODOURS ON YOUR HANDS

A great way to get rid of the smell of fish on your hands is to rub them with bicarbonate of soda before washing them.

DEODORISING CARPETS IN THE CAR

Our car used to smell of old dog, even though we didn't have one (a dog, that is!). We had a mystery problem with a leaky seal (no, not the kind that can perform circus tricks!). Putting bicarbonate of soda into the unused ashtray would have solved the problem. Alas, the car was written off in an accident before I could put it to the test. Otherwise, I would also have tried sprinkling bicarb onto the carpets before vacuuming.

DEODORISING TERRY NAPPIES

I know that most parents seem to use disposable nappies nowadays, but they weren't an option when my kids were babies. Soaking nappies in a bucket with a solution of bicarbonate of soda and water overnight was a ritual that not only got rid of the smell of ammonia but also cut down the amount of washing powder needed to keep them white and stainless. I can't recall the children ever suffering from nappy rash either.

Personal Uses for Bicarbonate of Soda

My research has led me to some interesting solutions to old problems as well as some truly bizarre ones. I've put in some of the old, natural remedies for interest rather than to recommend them. Obviously these are suggestions only and if you have any doubts, consult your doctor or a specialist in the field before launching in.

BATH BOMBS
Did you know that the main ingredient in fizzy bath products is bicarbonate of soda? You can make your own with three parts bicarbonate of soda to one part citric acid. Add your own fragrance oils or colouring. You can buy moulds for this purpose, but I think egg boxes would work just as well.

Mix together the dry ingredients and add water-based colour (not pigment, as this may stain). Don't be tempted to overdo the colour as this will be stronger when mixed with water. Mix with the oil of your choice (lavender is good for relaxation) until it clumps together in your hand. If the mixture is too stiff, add a little base oil, like almond oil. Put small balls of this into the egg box to dry. This will take a day or two. When you want to use one, just put in a bath of water and enjoy.

BATH DETOX
Use one part bicarbonate of soda to one part sea salt in a hot bath. Lie back and relax for 20 minutes. Rinse off or shower to remove the salt.

BATHTIME TREAT
Mix together four tablespoons of bicarbonate of soda, two of sugar, one teaspoon of ground cinnamon, a pinch of ground ginger and a pinch of ground cloves. This will keep for some time. When you are ready for a bath, take two tablespoons of the dry mix and add to running water.

BATHWATER SOFTENER
Use half a cup of bicarbonate of soda in the bath, and add to running water.

BEE STINGS
Mix a paste of bicarbonate of soda and water. Apply to the swelling, after the sting has been removed.

BRUSHES AND COMBS
These can be cleaned in a solution of bicarbonate of soda and warm water to remove build-up of products. If you can, immerse the whole brush. If you are bothered about harming the handle of a top of the range model or an antique, bristle variety, you can swish the bristles around in the water without immersing the whole thing. This was how the Victorians washed pure bristle brushes.

CLEANING CONTACT LENSES
Apparently, bicarbonate of soda in water cleans contact lenses. You must rinse very thoroughly, however, to avoid stinging.

CLEANING SKIN ABRASIONS
Wash with a mild solution of bicarbonate of soda and warm water.That sounds fine, but what about sprinkling the graze with black pepper while still wet? I leave that part of the remedy entirely up to you.

CLEANING SMELLS FROM YOUR HANDS
If you have been painting, gutting fish or preparing onions, you might like to get rid of the lingering odour with a sprinkling of bicarbonate of soda on your hands. Rub them together before washing with soapy water.

COLD SORES
Apply the bicarbonate of soda as a paste with water onto the sore, but don't rub. This will promote healing as well as soothing the affected area.

CURLING IRONS
Baked-on hair products, such as hairspray and gel, can be removed with a mixture of one tablespoon of bicarbonate of soda and a teaspoon of salt, mixed with a little vinegar. Turn off the tongs first, though!

CYSTITIS

Cystitis occurs when the bladder lining becomes inflamed and can mean you experience pain when passing water. There may be an urgent and frequent need to 'go' without producing much urine. This can be alleviated by a bath with bicarbonate of soda added, but don't use other oils or irritants. As well as drinking plenty of water to flush out the urinary system and drinking a glass of cranberry juice a day as a preventative, you could also try drinking a large glass of water with a teaspoon of bicarbonate of soda to make the urine less acidic. First and foremost, however, consider seeking the advice of your doctor.

DEEP CLEANSER

Mix an egg yolk with a teaspoon of vinegar, a teaspoon of lemon juice and a tablespoon of olive oil. Add a teaspoon of bicarbonate of soda. Apply to the skin with circular movements. Rinse off with warm water.

DRY SHAMPOO 1

I remember someone at school doing this with talcum powder when I was a teenager and everyone was obsessed with not having greasy hair. I can't think of when there isn't time to wash your hair nowadays in the shower, apart from when camping and/or faced with a lack of water. However, the tip is: sprinkle bicarbonate of soda onto your hair and rub through. Comb out and dry in warm air. Luckily I don't have greasy hair anymore!

DRY SHAMPOO 2

This sounds even less pleasant. Sprinkle one teaspoon of oatmeal and one of bicarbonate of soda on your head. Massage in for a few minutes, then comb or brush out. Again, warm air will help to get rid of residue.

DRY SKIN TREATMENT

Make a paste of bicarbonate of soda and olive oil. Apply to dry skin, for example rough elbows, and leave for three minutes. Rinse with warm water.

FACE WASH

A mild solution of bicarbonate of soda in warm water is good for washing your face without soap. It will also help to clean the pores and may prevent acne.

FOOT ODOUR EATER
Use bicarbonate of soda as a substitute for talcum powder after washing your feet. See below for a foot spa suggestion.

FOOT SPA
To put life and harmony back into tired, smelly feet, soak them in a bowl of hot water with four tablespoons of bicarbonate of soda. Bliss! The soaking will also soften hard skin, which you can remove with a pumice stone.

GARGLING AND MOUTHWASH
Gargling with half a teaspoon of bicarbonate of soda in a small glass of water will freshen and soothe your mouth and throat.

GUM MASSAGE
Slightly more astringent, but a powerful gum massage: take equal amounts of bicarbonate of soda and salt. Rub this into your gums with your finger.

Alternatively, sprinkle half a teaspoon of bicarbonate of soda onto your toothbrush and brush as normal.

HAND TREATMENT
To soften hands put two tablespoons of bicarbonate of soda into a bowl of hot water. Sit and soak your hands for about 20 minutes. Dry them and apply hand cream, then place your hands inside clean cotton gloves or clean socks for another 20 minutes and relax. Of course if you use the socks, you might have a bit of explaining to do, which may also cause you to laugh a lot, but laughing is good for you, too! In which case, don't try the next item at the same time.

LIP EXFOLIATOR
Mix a little bicarbonate of soda with a few drops of lemon juice. Work over your lips. Remove and cover with lip balm.

MILD ACNE
Use a paste of bicarbonate of soda with a drop of water. Apply to spots and leave for about ten minutes. Rinse off carefully with warm water but never rub or scrub your skin.

NAPPY RASH
As well as appreciating your use of bicarbonate of soda in the washing, your baby may benefit from a tablespoon in the bathwater to sooth nappy rash.

PERFORMANCE ENHANCER
Bicarbonate of soda has been used by athletes and cyclists to counter the build-up of lactic acid during races of up to 800 m or in time trials. There is also the risk of stomach upsets and diarrhoea, however, which might have the opposite effect!

RAZOR BURN
Dab a mild solution of bicarbonate of soda onto the affected areas. This works well after shaving armpits and acts as a mild deodorant as well.

RELIEVING ACID INDIGESTION
This age-old remedy has given relief to generations. Bicarbonate of soda is used in the manufacture of several commercial brands, including Alka-Seltzer and Andrews Liver Salts.

CAUTION

However, there may be several causes for prolonged indigestion, so you should consult your doctor if the symptoms persist. Also, if you are taking medication, you should check that the bicarbonate of soda does not interfere with other treatment and, if you suffer from heart problems or high blood pressure, you need to limit your intake of sodium. Obviously this remedy would be adding extra sodium. Taking bicarbonate of soda for indigestion should be considered as a temporary solution or for an occasional problem and you should never take indigestion remedies for more than two weeks at a time.

Children under six should never be given bicarbonate of soda for indigestion.

The normal dose for adults is half a teaspoon in water every two hours. Avoid taking large quantities of milk or milk products. Children from 6 to 12 years can be given quarter to half a teaspoonful in water after meals.

RELIEVING ACIDITY IN URINE

As already stated above, this is not a long term solution. Bicarbonate of soda makes urine less acidic. Uric acid can lead to kidney stone formation, so check with your GP if in doubt.

Take one teaspoon in water every four hours. Do not take more than four doses a day. Remember also that this will increase your sodium intake a lot, so beware if you have high blood pressure.

RELIEVING INSECT BITES

Apply a paste of bicarbonate of soda and water to the affected area. This should relieve the irritation or discomfort.

RELIEVING MILD SKIN INFECTIONS

Some of these suggestions date back centuries. I offer them for interest.

Make a poultice as follows. Slice two onions finely and cook until tender in a pint of water. Thicken with wheat bran and add half a tablespoon of bicarbonate of soda. Put this poultice on the affected area and change as often as is necessary to reduce inflammation.

OR, HOW ABOUT A PANCAKE REMEDY?

Make a pancake with sour milk, bicarbonate of soda and wheat flour. Cover the affected part and keep the pancakes coming so that they stay warm. Maybe not!

RELIEVING SKIN IRRITATION

If you are suffering from a skin irritation or are getting over chickenpox, try adding three tablespoons of bicarbonate of soda to a warm bath.

REMOVING A BUILD-UP OF HAIR PRODUCTS

If you have been using a lot of hair products, you can get rid of the build-up by mixing two tablespoons of bicarbonate of soda with a tablespoon of shampoo. Massage this into your hair while you are washing it and rinse thoroughly. Don't put anything else on straightaway, like conditioner, which you probably don't need anyway.

REMOVING SPLINTERS

Mix a little water with half a teaspoon of bicarbonate of soda to form a paste. Apply this to a splinter and cover it with a plaster for a couple of hours. The soda should draw out the splinter, which can then be easily removed, along with any toxins, and solve the problem.

STOMACH ACHE

Mix together one teaspoon of bicarbonate of soda with three tablespoons of cider vinegar. After the effervescence has died down fill the glass with cold water and drink the whole glass full. Relief should follow rapidly.

SUNBURN

Relieve sunburn by making a paste of bicarbonate of soda and water and spreading over the affected area. Alternatively, put a cup of bicarbonate of soda into a tepid bath for relief and soothing.

TOOTH WHITENER AND CLEANER

After brushing with toothpaste, apply a little bicarbonate of soda to the toothbrush and re-brush your teeth. This will help to whiten them.

TOOTHBRUSH CARE

This may be useful when camping or travelling. Clean your toothbrush with a little sprinkle of bicarbonate of soda and rinse in water.

UNDERARM DEODORANT

Dust the armpits with bicarbonate of soda after washing or showering.

WART TREATMENT

Apply a few drops of vinegar to the wart and dust with bicarbonate of soda. Remove after 15 minutes. Repeat several times a day, until the wart disappears.

Cooking with Bicarbonate of Soda

One of the main advantages of using bicarbonate of soda in baking is that it is easy to use as a raising agent. Bakery products are aerated by gas bubbles, developed either naturally or by being folded in from the atmosphere. Bubbles can be achieved by a yeast or bacterial fermentation, from chemical reaction induced by combining, for example, bicarbonate of soda with another ingredient, or by injecting gas through whisking or beating. All commercially produced cereal breads, except rye bread and salt-rising varieties (soda bread), use yeast. Cakes and biscuits are usually leavened by carbon dioxide which is produced when bicarbonate of soda, also commonly called baking soda, is added. When it is added on its own it makes dough alkaline and causes deterioration of flavour and colour. For these reasons an acid-reacting substance is also included.

CREAM OF TARTAR

There has been a lot of research over the years into improving and stabilising the above effects, since the early days of using soda ash for baking. Leavening acids regulate the rate of gas (carbon dioxide) released. The rate of release affects the size of the bubbles and influences the texture and volume of the cooked product. Acetic acid (from vinegar) and sour milk usually react quickly with bicarbonate of soda (see the Fun with Bicarbonate of Soda section). Cream of tartar (potassium acid tartrate), sodium aluminium sulphate (alum) and calcium phosphate have all been added, but cream of tartar is the most commonly known and widely used leavening acid.

WHAT A GAS

Instead of using bicarbonate of soda, then, most commercial companies use baking powder, which is a mix of soda and acids in the right amounts, with additives such as cornflour, which make the product easier to measure and improve stability. So, baking powder is simply a mixture of carbon dioxide and some harmless salts.

WHAT'S THE DIFFERENCE?

So when should you add baking powder, as opposed to bicarbonate of soda? If you use bicarbonate of soda the batter or mix needs to include an acid such as that found in honey, brown sugar, yoghurt, fruit or cocoa. The reaction that takes place will be fast, requiring you to work quickly and start cooking the ingredients as soon as possible, i.e. before the carbon dioxide has finished fizzing.

Baking powder already has the acid content needed to form a chemical reaction in the cream of tartar, so only needs moisture, usually in the form of milk or water, as a catalyst to start the reaction.

DOUBLE BUBBLE

Most commercial baking powders are double-acting, giving off small amounts of carbon dioxide during the mixing and making stages, then staying inactive until baking raises the temperature of the batter. This prevents a great deal of loss of gas which may come about if the batter is left waiting to be cooked for a long time. If the ingredients need a lot of blending, the double-action effect is obviously going to work better.

Just in case you are still interested and haven't given up this section yet in favour of reading the recipes, the other way to leaven food and produce the desired effect is by trapping air through beating, folding and whisking. Sponge cakes, for example, can be prepared without either yeast or baking powder. Pockets of air are folded in during the beating of eggs. A foaming ingredient like egg white is the ideal ingredient for this purpose.

I now understand why, when using a food processor to make a cake mix, you need to add a raising agent to help lift the batter; there isn't enough whisking going on to get the eggs fluffy and incorporate air into the mix.

Another interesting fact I've learned through my research is that water vapour exerts quite a lot of pressure on the inside walls of bubbles already formed, as the cake or loaf nears boiling point. This water can obviously have quite an effect on whether the cake sinks or stays light (and explains why my Yorkshire puddings always collapse!). If you add too much baking powder or bicarbonate of soda to a mix, in the hope of giving it an extra rise, you risk the possibility of the bubbles bursting and the cake sinking into the sunset. It will also give the cake a bitter taste. Equally, too little bicarbonate of soda or baking powder results in a tough, rubbery end product.

SELF-RAISING FLOUR
This is basically plain flour which has had small amounts of baking powder added during manufacture.

COOKING USES AND TIPS

Please remember that baking powder and bicarbonate of soda are different chemical compounds, so you can't use one in place of the other for the reasons explained above. The following hints and tips for using bicarbonate of soda are included as interesting ideas rather than recommendations.

MAKE YOUR OWN BAKING POWDER
You can make your own baking powder: for every teaspoon of baking powder, substitute

½ teaspoon cream of tartar
⅓ teaspoon bicarbonate of soda
A pinch of salt

Or

¼ teaspoon bicarbonate of soda,
110 ml (4 fl oz) buttermilk or yoghurt

Or

¼ teaspoon bicarbonate of soda,
¼ teaspoon cornflour
½ teaspoon cream of tartar

NORMAL AMOUNTS TO USE
The normal amount of bicarbonate of soda to add to each mixture is half to one level teaspoon per 240 ml (8 fl oz) liquid.

For baking powder use one teaspoon for every 150g (5oz) flour.

COOKING AT HIGH ALTITUDE

If you live at high altitude you can expect carbon dioxide gas to expand at a faster rate and so it will have greater leavening, but you always need at least half a teaspoon of bicarbonate of soda for every 240 ml (8 fl oz) of liquid.

TEST FOR FRESHNESS (SODA)

To check if bicarbonate of soda is still viable, mix a quarter of a teaspoon with two teaspoons of vinegar. Look for an instant fizz. Soda has an indefinite shelf-life, but should be stored in an airtight container in a cool, dry place.

TEST FOR FRESHNESS (BAKING POWDER)

Mix one teaspoon with 120 ml (4 fl oz) of hot water. There should be an instant reaction. The shelf-life of baking soda is shorter than that of bicarbonate of soda, so you should take note of the sell-by date on the carton.

SUBSTITUTE FOR EGGS

If you are short of eggs or have an allergy or other reason for not eating them, replace each egg with half a teaspoon of baking powder and two tablespoons of liquid.

SUBSTITUTE FOR MILK IN SCRAMBLED EGG

Instead of adding milk to make eggs go further, add a half teaspoon of bicarbonate of soda per egg to make light and fluffy scrambled eggs. I suppose this may be useful to anyone with a milk allergy, but I think I'd rather stick with an omelette in that case.

BATTERS FOR FRYING

The addition of bicarbonate of soda or baking powder to some types of batter makes them very light and crispy. Look for some recipes in a later section.

OTHER (WO)MAN'S GREENS ALWAYS GREENER?

The addition of bicarbonate of soda to water when cooking greens such as cabbage used to be common, but research has shown that this practice destroys the vitamins B1 and C, so is not recommended. It's far better to steam them, so that you don't lose the colour in the cabbage water.

PRETZELS

Bicarbonate of soda is used in the making of pretzels to give them their dark brown appearance.

TOMATO SAUCE

Bicarbonate of soda can be added to tomatoes to neutralise their acidity when you are making tomato sauce.

WINDS OF CHANGE?

If you add bicarbonate of soda to the water when cooking beans, it will avoid flatulence once you have eaten them. Alternatively, soak dried beans in water with added bicarbonate of soda to aid digestion . . . and avoid curtain raisers!

TENDERISING MEAT

You can add a teaspoon of bicarbonate of soda to beef stew, to speed up the cooking time and break down the tough fibres, apparently. Alternatively, you could just cook it for longer. Some restaurants allegedly tenderise tougher cuts of beef and pork by soaking the meat in a solution of bicarbonate of soda and water and leaving it for several hours in the refrigerator. Later, the meat is rinsed thoroughly to get rid of the soda residue.

PLUCKING A CHICKEN

I have no experience of this, but apparently if you add a teaspoon of bicarbonate of soda to boiling water to scald a dead chicken or other feathered game, the feathers will come off more easily.

REMOVING A 'HIGH' SMELL

The odour and intensity of wild game can be tamed by washing the bird in water with bicarbonate of soda.

CRISPING MEAT. . .

. . . or rather, crisping the fat on pork chops or chicken cooked in the skin. This is not something I can recommend because I don't eat pork fat or chicken skin, but apparently you can crisp the fat by rubbing bicarbonate of soda into the skin before cooking.

COOKING RHUBARB
When simmering or boiling fresh rhubarb, add a teaspoon of bicarbonate of soda to neutralise the acid. It will decrease the amount of sugar required to sweeten the batch. The appearance of the rhubarb will change slightly.

SAVOURY RECIPES

As well as having obvious properties in the leavening of cakes and bread bicarbonate of soda brings a lightness to batters and fried foods, especially in vegetarian recipes. Many different styles of cuisine throughout the world are represented below in some adapted recipes.

TEMPURA
One of the most popular ways of using baking powder seems to be as an ingredient in tempura, a classic Japanese method of deep frying seafood and vegetables. The Japanese didn't invent the process, however: it was introduced by Portuguese missionaries in the 16th century. Some versions of tempura use eggs, but the versions below don't. Opinion is divided over which is more traditional. All manner of seafood, fish and vegetables are prepared this way and are dipped into sauce or lightly salted before eating.

The main difference between tempura and other batters is the texture. Tempura batter is not like pancake batter in texture. Normal batter is beaten until very smooth, but tempura still has lumps in it when cooked. Iced water is always used, which slows down the rate of expansion of the batter until it reaches the hot oil. Normally, when you put water with flour it starts to expand straight away, but the temperature slows this process.

There should be dry spots in the batter, which will give a light, crisp coating. The tempura doesn't need to go brown. The whole idea is to encase the food in a light batter which is crispy but not saturated in oil. The technique is to lower the battered food into the hot oil so that it doesn't stick to the bottom of the pan.

Tempura is served as part of a meal or as a snack. The most popular seem to be filled with shrimp, prawn, squid or other fish, shiitake mushrooms, sweet potato, tofu or firm vegetables like green peppers, aubergines, courgettes, but there are sweet versions as well, whereby you can have tempura ice cream with bananas or maraschino cherries.

TEMPURA BATTER

You will need:

80 g (3 oz) plain white flour
125 g (4½ oz) cornflour
2 tablespoons baking powder
Salt and freshly ground black pepper
270 ml (9 fl oz) chilled water

Method:

1. Put the flours, baking powder, salt and pepper into a large bowl. Add the chilled water but don't overmix. Lumps are fine.

AUBERGINE TEMPURA

You will need:

Quantity of basic tempura batter
2 aubergines
Salt
Vegetable oil for deep frying

Method:

1. Wash the aubergines then cut in slices about 1 cm (½ in) thick. Place in layers in a colander, sprinkling each layer with a little salt. Cover with a plate and leave for 15 minutes.
2. Prepare a quantity of tempura, as above.

3. Rinse the aubergine slices under cold running water and dry well. Coat with a little flour then dip into the batter.
4. Heat the deep fryer, or oil in a pan or wok to 190°C (375°F). Lower the slices in, making sure that they do not stick to the bottom of the pan. Deep fry in batches, for 1 to 2 minutes only, until crispy but not browned. Drain on kitchen paper. Serve immediately with a dipping sauce of your choice.

COURGETTE TEMPURA

You will need:

4 courgettes (zucchini)
Quantity of tempura batter
Vegetable oil for deep frying

Method:
1. Wash the courgettes and cut into batons 1 cm (½ in) thick and 5 cm (2 in) long. Dry well on kitchen towels.
2. Prepare a quantity of tempura batter, as above.
3. Dust the courgette batons with a little cornflour to coat, then dip into the batter.
4. Heat the deep fryer, wok or pan of oil to 190°C (375°F). Gently lower the batons in so that they do not stick to the bottom, or each other. Remove with a slotted spoon when crisp, but before they brown. Serve straight away.

OTHER VEGETABLE SUGGESTIONS FOR TEMPURA

PEPPERS
Prepare red or green peppers by washing, removing the seeds and slicing through from top to bottom. Divide each half into two or three pieces.

ONIONS
Use large onions. Remove outer skin, cut in half from top to bottom. Slice each half into 1 cm (½ in) slices and secure each slice with a cocktail stick inserted from the top of the arc.

TOFU
Slice to 2 cm (1 in) thick, then slice into sticks.

MUSHROOMS
Use whole, or if very fat, slice through the centre.

DIPPING SAUCE SUGGESTIONS

SWEET-AND-SOUR SAUCE

You will need:

- 120 ml (4 fl oz) honey
- 1 tablespoon cornflour
- 80 ml (3 fl oz) red wine vinegar
- 80 ml (3 fl oz) chicken stock
- 1 green pepper, finely chopped
- 1 tablespoon soy sauce
- ¼ teaspoon garlic powder
- ¼ teaspoon ground ginger

Method:
In a saucepan, combine honey and cornflour. Stir in the vinegar, chicken stock and the other ingredients. Simmer for a few minutes. Serve hot or cold.

ORIENTAL DIPPING SAUCE

You will need:

- 120 ml (4 fl oz) honey
- 60 g (2 oz) peanut butter
- 1 tablespoon soy sauce
- 1 clove of garlic, crushed
- 1 teaspoon freshly chopped coriander

Method:
Place all the ingredients in a bowl and mix well.

CHICKEN TEMPURA

This chicken in batter may appeal if you don't like the spicy, Indian pakora version on page (82).

You will need: **Serves 4 to 6**

Quantity of basic tempura batter, as above
Vegetable oil for deep frying
4 chicken breasts, cut into thin strips

Method:

1. Prepare a quantity of tempura batter, as above.
2. Dust the chicken gougons in a little extra cornflour then dip into the batter.
3. Preheat a deep fryer to 190°C (375°F), or prepare oil in a large saucepan. Fry the chicken in batches for 2 to 3 minutes, until golden brown. Drain on kitchen paper and serve immediately.

OTHER SAVOURY RECIPES

POTATO GNOCCHI

A recipe for using up leftover mashed potato. This recipe serves six, so you might need to scale it down a little.

You will need:

675 g (1½ lb) potatoes, peeled and cut into chunks
150 g (5 oz) wholemeal flour
½ teaspoon baking powder
1½ tablespoons olive oil
Salt and black pepper

Method:

1. Boil the potatoes in a pan of water for about 20 minutes, until soft.
2. Drain well and mash until smooth. Add the remaining ingredients and mix well to form a dough, using extra flour if sticky.
3. Take small pieces and roll into balls. Flatten them with a fork and use the prongs to make ridges. Fold these in half with the ridges outside.
4. Add the gnocchi to a large pan of salted water and cook until they float to the top. This will be about 2 minutes, but larger versions will take a bit longer. Drain and serve with a sauce of your choice.

EGYPTIAN BEAN CAKES

Broad beans are a member of the pea family. If you use dried beans for this recipe you need to soak them overnight. This is a good recipe for gardeners with a glut of broad beans. The shelling and, after cooking, the skinning of them is best done as a social activity while chatting or sunbathing.

You will need: **Serves 4 to 6**

400 g (14 oz) shelled broad beans
4 tablespoons fresh dill leaves (2 tablespoons dried)
4 tablespoons chopped coriander
4 tablespoons parsley
2 onions, chopped

10 cloves garlic
1 teaspoon ground cumin
½ teaspoon cayenne pepper
1 teaspoon bicarbonate of soda
Salt
2 tablespoons sesame seeds
Vegetable oil for frying

Method:

1. Drain the pre-soaked beans, if they were dried. Remove the skins, then place in a food processor together with the herbs, spices and bicarbonate of soda and process to a rough paste.
2. Transfer to a large mixing bowl and knead for a few minutes. Cover and leave to stand for half an hour.
3. Scoop small amounts of the mixture and shape into cakes about 5 cm (2 in) across. Sprinkle one side with sesame seeds.
4. Put some oil into a large frying pan. When hot, fry the cakes for a few minutes until golden brown. Don't overcrowd the pan. Drain on kitchen paper and serve hot or cold.

CHESTNUT FRITTERS

These can be served to accompany roast turkey for Christmas, but are equally good for party snacks or as a starter with a green herb salad and some dressing. Instead of frying you could cook them in the oven on a small baking tray while the meat is cooking. They will take about 15 to 20 minutes. Turn halfway through.

You will need:

400 g (14 oz) tin puréed cooked chestnuts
50 g (2 oz) plain flour
½ teaspoon baking powder
3 or 4 bacon rashers, chopped

Method:

1. Mix all the ingredients together until well blended.
2. Form into balls and flatten them slightly. Make small balls for party food, or eight larger ones for a starter. You may need to flour your hands.

3. Fry for a few minutes each side, until golden. Serve hot.

BATTER FOR FISH

You will need:

80 g (3 oz) plain flour
2 tablespoons cornflour
¼ teaspoon bicarbonate of soda
¼ teaspoon baking powder
¼ teaspoon salt
80 ml (3 fl oz) water

Method:

1. Sift the dry ingredients together.
2. Add the water and mix well.
3. Place in a flat dish and coat the fish completely.
4. Deep fry until a nice golden brown.

FALAFEL

This recipe uses the bicarbonate of soda to reduce flatulence, as described in a previous section. It needs overnight soaking to be effective.

You will need: **Makes about 16**

200 g (7 oz) dried, peeled and split broad beans
110 g (4 oz) dried chickpeas
1 teaspoon bicarbonate of soda
4 garlic cloves
1 onion, roughly chopped
1 small leek, sliced
3 tablespoons fresh, chopped coriander
½ teaspoon bicarbonate of soda
1 teaspoon ground cumin
1 teaspoon ground allspice
Pinch of cayenne pepper
Salt and freshly ground black pepper

Method:

1. Soak the broad beans and chickpeas in separate bowls overnight, stirring in half of the bicarbonate of soda into each bowl.
2. Skin the chickpeas, and drain and rinse both pulses in cold water.
3. Place in a blender with other ingredients to produce a smooth paste. Allow this mix to rest for 30 minutes.
4. Shape the mixture with your hands into round cakes of 5 to 6 cm (2 in) across and 2 cm (¾ in) thick.
5. Heat some vegetable oil in a large pan and when hot fry the falafel until golden brown on both sides. Drain on several layers of kitchen paper and serve hot or cold, with pitta bread and tomatoes or pickled cucumbers.

TOMATO, BASIL AND CHEESE SKEWERS

These little cheesy tomato skewers use mozzarella, although you could experiment with other varieties, such as goat's cheese or even a harder cheese. They will make an interesting starter or another unusual party snack. Once the preparation is done, they only take minutes to cook.

You will need: **Serves 4**

12 cherry tomatoes, halved
12 large basil leaves
2 mozzarella cheeses
Vegetable oil for frying

For the batter:

110 g (4 oz) rice flour
110 g (4 oz) plain flour
2 heaped teaspoons bicarbonate of soda
150 ml (5 fl oz) soda water (maximum)

Method:

1. Pick the largest basil leaves. Push one end of a leaf (close to the end) onto a small skewer or cocktail stick. Ideally, later you will push the other end of the leaf onto the stick.
2. Push half a tomato onto the stick, followed by a small

piece of cheese so that it touches the tomato. Push another half tomato on so that the flesh side of the tomato touches the cheese, then push the other end of the basil leaf onto the stick. Put these into the fridge until you are ready to cook.

3. Sift the rice flour, plain flour and bicarbonate of soda into a bowl. Add 100 ml of the soda water and whisk thoroughly. Add a little more of the soda water until you have a smooth batter.

4. Heat the oil in a large saucepan or pot. If you put a cocktail stick in it should sizzle. Give the cheese and tomato sticks a good even coating of batter and drop straight into the oil with tongs or a slotted spoon for 45 seconds to a minute. This should be long enough to crispen the batter but not long enough to melt the cheese.

Serve as they are or with a dip or salad.

BASIC PAKORAS

I love cooking Asian food and once you have assembled all the spices it is easy to produce authentic dishes whenever you want. This batter recipe can be used to give any meat or vegetables a spicy coating, but I have included one for chicken below.

You will need: **Serves 4 to 6**

200 g (7 oz) chickpea flour
½ teaspoon salt
1 teaspoon bicarbonate of soda
½ teaspoon chilli powder
1 teaspoon ground coriander
1 teaspoon garam masala (mix of spices for curry)
Approx. 300 ml (10 fl oz) water

You will need for CHICKEN PAKORAS:

450 g (1 lb) boneless chicken breast
1 tablespoon lemon juice
1 teaspoon crushed garlic
1 teaspoon fresh ginger, grated
½ green chilli, finely chopped

½ teaspoon ground turmeric
1 teaspoon ground cumin
½ teaspoon salt
1 teaspoon ground coriander
Oil for deep frying
2 tablespoons freshly chopped coriander for garnish

POTATO, ONION AND SPINACH PAKORAS

You can use the basic recipe but substitute the following for chicken and you won't need to marinade it for 2 hours.

You will need:

175 g (6 oz) grated potato
1 large, finely chopped onion
110 g (4 oz) washed, chopped spinach
2 green chillies, finely chopped

Method:

Preparing the batter

1. Put the dry ingredients for the batter into a large bowl and make a well in the centre. Gradually add the water to make a smooth, thick batter. Leave aside for about 20 minutes.

Chicken recipe

If you are using chicken it should be marinated for at least 2 hours so that the meat absorbs plenty of the flavour from the spices, so you should complete this preparation before making the batter.

1. Cut the chicken into bite-sized pieces. Place in a bowl and rub with the lemon juice.
2. In a small mixing bowl, mix the garlic, ginger, green chillies, salt, turmeric, cumin and ground coriander. Pour over the chicken pieces to coat well. Cover and set aside for at least 2 hours.

Cooking the battered food

1. When ready to cook, heat the oil to 180°C (350°F).
2. Dip a few pieces of chicken into the batter to thoroughly coat, then, using tongs, carefully drop them into the hot oil. Fry for a few minutes, then turn them

over and continue to fry for a further 5 to 8 minutes until crisp and golden.

3. Remove with a slotted spoon and drain on kitchen paper. Repeat with the remaining chicken pieces. Serve hot or cold, garnished with fresh coriander.

INDIAN CHICKPEA DUMPLINGS

You will need:

100 g (4 oz) chickpea flour
Large pinch of salt
½ teaspoon bicarbonate of soda
125 ml (4 fl oz) plain yoghurt

Method:

1. Sift together the flour, salt and bicarbonate of soda.
2. Add the yoghurt and mix well with a wooden spoon to make a thick paste. Continue to beat for about 5 minutes, until light and airy.
3. Drop spoonfuls into a pan of hot oil and fry until golden brown. Drain on kitchen paper.
4. To serve, add to a simmering soup or vegetable dish about 10 minutes before serving.

VEGETABLE FRITTERS

You will need: **Serves 4**

250 g (9 oz) chickpea flour
50 g (2 oz) self-raising flour
3 tablespoons cornflour
1 teaspoon baking powder
1 large potato, peeled and grated
1 large carrot, grated
1 large onion, finely chopped
¼ teaspoon chilli powder (optional)
¼ teaspoon ground cumin
¼ teaspoon ground coriander
¼ teaspoon ground turmeric
1 tablespoon oil
1 tablespoon lemon juice
Approx. 240 ml (8 fl oz) water
Oil for frying

Method:

1. Sift the three flours and baking powder into a large mixing bowl.
2. Add the vegetables, spices, lemon juice and oil to the flour mixture, plus enough water to make a thick batter. Leave to stand for 5 minutes.
3. Heat some oil in a large frying pan or fryer. Drop tablespoons of the batter into the oil and fry on both sides until golden brown. These will probably need to be done in batches. Drain on absorbent paper and serve hot.

CARIBBEAN SALTFISH CAKES

These fish cakes are called Stamp and Go; I don't know why. When using salted fish you need to soak it in advance for 2 to 4 hours.

You will need: **Makes 24**

225 g (8 oz) salt fish
225 g (8 oz) plain flour
2 teaspoons baking powder
½ teaspoon salt
Freshly ground black pepper
1 egg, lightly beaten
60 ml (2 fl oz) milk
1 onion, finely chopped
1 garlic clove, crushed
1 fresh chilli pepper, deseeded and finely chopped
Vegetable oil for shallow frying

Method:

1. Drain the pre-soaked fish, rinse and place in a saucepan. Cover with fresh water and bring to the boil. Reduce the heat, cover and simmer for 10 minutes, or until cooked.
2. Drain the fish again, but this time reserve the liquid. Allow to cool, then remove skin and bones and flake the flesh. Set aside.
3. Sift the flour, baking powder and salt together into a large mixing bowl.
4. In another bowl, combine the egg, milk and 60 ml (2 fl oz) of the reserved liquid, then stir into the flour

mixture a little at a time to make a thick batter.
5. Add the flaked fish, onion and pepper and mix well.
6. Preheat the oil to 180°C (350°F) in a large frying pan then drop tablespoonfuls of the fish mixture into the hot oil, a few at a time, and fry for 2 to 3 minutes until golden brown. Drain on paper towels and serve hot or cold.

ONION BHAJIS

If you don't like your bhajis too hot, leave out the fresh chilli.

You will need: **Makes 10 to 12**

110 g (4 oz) chickpea flour
¼ teaspoon chilli powder
½ teaspoon turmeric
½ teaspoon baking powder
½ teaspoon ground cumin
1 large onion, thinly sliced
1 green chilli, deseeded and finely chopped
25 g (1 oz) finely chopped fresh coriander
Salt
Cold water to mix
Vegetable oil for deep frying

Method:
1. Sift the flour, chilli, turmeric, cumin, baking powder and salt into a large mixing bowl. Add the chopped coriander, onions and chillies (if using) and mix well.
2. Gradually add just enough water to the flour mixture to form a thick batter. Mix well so that the onions are well coated.
3. Preheat a deep fat fryer or wok to 180°C (350°F). Using a slotted spoon, drop spoonfuls of the mixture into the hot oil and fry until brown and crispy. Do not overcrowd the pan.
4. Keep warm while you cook the remaining bhajis. Drain and serve hot.

TOFU AND POTATO PASTIES

These are good for packed lunches or picnics.

You will need for the pastry: **Serves 4 to 6**

- 275 g (10 oz) mashed potato
- 175 g (6 oz) plain flour
- 1 egg
- 25 g (1 oz) oil
- ½ teaspoon baking powder
- Salt and freshly ground black pepper
- Milk for mixing and glazing

You will need for the filling:

- 150 g (5 oz) tofu, mashed
- 2 carrots, grated
- 1 onion, finely chopped
- 110 g (4 oz) spring cabbage, shredded
- 1 clove garlic, crushed
- 1 tablespoon olive oil
- 1 teaspoon fresh ginger, grated
- ½ teaspoon ground coriander
- ½ teaspoon black mustard seeds
- ½ teaspoon chilli powder
- Salt and pepper

Method:

1. Put the potato, oil, salt and pepper into a bowl and mix well.
2. Sift the flour and baking powder into the bowl and mix with the egg and a little milk. Use just enough to make a firm dough.
3. For the filling, heat the oil in a pan and add the garlic and ginger. After a minute add the onion, mustard seeds, coriander and chilli powder. Brown the onions and then add the carrot, cabbage, tofu and seasoning. Cook for 3 or 4 minutes, until vegetables are just turning soft.
4. Preheat the oven to 200°C (400°F, gas mark 6) and grease a baking sheet.
5. Divide the dough into four or six pieces and roll out to 17 cm (7 in) circles on a floured board. Divide the tofu mixture between the pasties, leaving the edges clean. Brush with water and pinch the edges together to make a raised ridge. Glaze with milk.

6. Place on the baking sheet and cook for 20 minutes, or until brown. Serve hot or cold.

DELIGHTFUL DESSERTS

These recipes fall into two categories: where the bicarbonate of soda is used to leaven cake-like desserts and is also used to lighten batters for fritters and pancakes.

LEMON DUMPLINGS

You will need: **Serves 4**

110 g (4 oz) plain flour
2 teaspoons baking powder
1 tablespoon melted butter
1 egg
Pinch of salt
Milk for mixing

You will need for the sauce:

110 ml (3½ fl oz) boiled water
1 tablespoon butter
175 g (6 oz) sugar
1 tablespoon golden syrup
Juice and grated zest of a lemon

Method:

1. Sift the flour, baking powder and salt together.
2. Add melted butter and lightly beaten egg to form a dough with a little milk.
3. Form into marble-sized balls.
4. Combine the sauce ingredients in a large saucepan and bring to the boil.
5. Drop the dumplings in. Cover and simmer for about 20 minutes.
6. Serve hot.

UPSIDE-DOWN GINGER AND FRUIT PUDDING

This can be made ahead and stored in an airtight container for the flavours to mature. When you want to eat it just warm through and serve with custard, cream or crème fraiche. You could substitute pears or pineapple for the apple.

You will need for the base/topping: **serves 6**

60 g (2 oz) soft margarine
80 g (3 oz) brown sugar
1 tablespoon fresh lemon juice
2 cooking apples, peeled, cored and sliced

You will need for the sticky gingerbread:

450 g (1 lb) plain flour
225 g (8 oz) dark brown sugar
175 g (6 oz) black treacle
175 g (6 oz) golden syrup
175 g (6 oz) margarine
½ teaspoon bicarbonate of soda
2 teaspoons baking powder
3 teaspoons ground ginger
½ teaspoon cinnamon
1 egg
225 ml (8 fl oz) warm milk

Method:

1. Preheat the oven to 170°C (325°F, gas mark 3). Grease and line a baking tin, approx. 30 x 20 x 5 cm (12 x 8 x 2 in).
2. Cream the margarine and sugar together with the lemon juice for the base. Spread this evenly over the base of the tin. Arrange the sliced fruit on top.
3. Sift the flour, bicarbonate of soda, baking powder, cinnamon and ginger together into a large bowl.
4. In a saucepan gently warm the sugar, syrup, treacle and margarine.
5. Add the lightly beaten egg to the milk and pour these into the flour mix. Beat and combine with the syrup in the saucepan. Mix thoroughly.
6. Pour the batter over the fruit in the prepared tin and bake for 1 hour, or until firm. Leave the cake to cool in the tin for about 30 minutes, so that the fruit sets. Once completely cool, turn out of the tin and remove the paper.

HOPPERS

These wafer-thin, cup-shaped pancakes come from Sri Lanka, where they are a popular breakfast food. They need plenty of preparation time to soak and ferment, so you need to plan well ahead. You also need a pan with a lid to let the hoppers cook properly. They should be crisp on the outside and spongy and soft in the centre, and can be eaten with curries as well as fruity, sweet desserts.

You will need: **Serves 4 to 6**

360 ml (12 fl oz) water
150 g (5 oz) plain flour
225 g (8 oz) long grain rice
2 tablespoons sugar
480 ml (16 fl oz) coconut milk
Pinch of salt
½ teaspoon bicarbonate of soda

Method:

1. Soak the rice in cold water overnight.
2. Drain, then transfer to a food processor together with the (measured) water. Blend to a paste.
3. Transfer to a large mixing bowl. Add the sugar and flour and mix to a very thick batter. Leave in a warm place for 8 hours.
4. Add the bicarbonate of soda and salt to the batter, mix well then gradually add the coconut milk. Mix well, but not too vigorously.
5. Lightly oil a pan or wok and place over a moderate heat. Pour in a ladle of the batter and quickly swirl until the pan is coated.
6. Cover the pan and leave for about 5 minutes or until there is a golden edge. Try not to open the lid until the hopper is ready as it may collapse. Loosen the edges and slide the hopper out. Repeat with the remaining batter.

CHOCOLATE BANANA CUSTARD CAKE

This is more of a banana custard trifle, really. It would be the ideal way to use up a broken cake, or to use up half a leftover cake. Does anyone ever have cake left over?

You will need:

275 g (10 oz) self-raising flour
1 teaspoon bicarbonate of soda
250 ml (8 fl oz) vegetable oil
225 g (8 oz) caster sugar
150 ml (5 fl oz) orange juice
150 ml (5 fl oz) water
570 ml (1 pt) ready-made custard, cooled
110 g (4 oz) plain chocolate

- 1 to 2 tablespoons milk
- 2 to 3 bananas

Method:

1. Preheat the oven to 170°C (325°F, gas mark 3). Grease a pair of 20 cm (8 in) sandwich tins, or one rectangular baking tin.
2. Sift the flour and bicarbonate of soda into a large bowl and add the sugar.
3. In another bowl, mix together the oil, orange juice and water.
4. Combine the wet and dry ingredients and mix well to a smooth, wet batter.
5. Pour into the prepared cake tins and bake for about 30 minutes, until the cakes start to shrink away from the sides of the tins. Turn out to cool on a wire rack.
6. To make the chocolate sauce melt the chocolate with the milk in a bowl over a pan of warm water.
7. To assemble, crumble the cake into a large dish. Add the cooled down custard, and cut the bananas into slices. Layer these on top of the custard and finally pour the chocolate sauce over to finish. Chill for 2 to 3 hours then serve.

STICKY TOFFEE CHRISTMAS PUDDING

Definitely a special occasion pudding. It can be made a day or two ahead and kept cool.

You will need:

- 175 g (6 oz) self-raising flour
- 150 g (5 oz) sugar
- 200 g (7 oz) dates, stoned and chopped
- 110 g (4 oz) walnut pieces
- 1 carrot, peeled and grated
- 1 dessert apple, peeled, cored and grated
- 2 tablespoons golden syrup
- 4 tablespoons brandy
- 110 g (4 oz) butter or margarine
- 2 eggs, beaten
- 1 teaspoon bicarbonate of soda

You will need for the walnut toffee sauce:

175 g (6 oz) sugar
25 g (1 oz) walnut pieces
110 g (4 oz) butter
6 tablespoons double cream

Method:

1. Preheat the oven to 170°C (325°F, gas mark 3). Grease and line the base of a 1.5 l (3 pt) pudding basin with greaseproof paper.
2. Put the butter and sugar into a large bowl and beat until light and fluffy. Add the eggs, one at a time, and then sift in the flour and bicarbonate of soda. Stir in the carrot, apple, nuts, syrup and brandy.
3. Tip into the pudding basin and bake for about 70 to 80 minutes.

To make the sauce:

1. Tip the sugar into a frying pan and put over a medium heat. Cook until bubbling and golden. Swirl in the walnuts and, when completely coated, use a slotted spoon to lift them out. Place on a tray lined with greaseproof paper and leave the praline to harden.
2. Carefully add the butter and cream to the caramel residue in the pan, return to the heat and stir together to make a toffee sauce.

To serve:

Break the praline into shards. Turn out the pudding, spoon over some sauce and decorate with the praline. Serve with the remaining sauce.

DRIED FRUIT PUDDING

This recipe reminds me of the steamed puddings my mother and grandmother used to make, before the days of microwaves and instant desserts and when mums stayed at home and baked, being what my daughter calls a 'proper mother'! All I can say is thank goodness I'm not a proper mother. I've included this recipe to add interest. I vividly remember seeing the tea towels or pudding cloths on the line after being washed for re-use. Fun, eh?

You will need:

175 g (6 oz) self-raising flour
175 g (6 oz) brown breadcrumbs
175 g (6 oz) melted margarine
1 teaspoon bicarbonate of soda
2 teaspoons ground cinnamon
1 teaspoon ground ginger
110 g (4 oz) currants
175 g (6 oz) sultanas
110 g (4 oz) soft dark brown sugar
2 tablespoons golden syrup
360 ml (12 fl oz) milk

Method:

1. Place a clean, cotton or linen tea towel in hot water to soak.
2. Place all the ingredients in a large mixing bowl, and beat together to form a fairly soft consistency.
3. Place a heatproof plate in the bottom of a large saucepan and add enough water to fill about a third of the pan. Bring to the boil.
4. Meanwhile, wring out the tea towel and spread on a surface. Dredge well with flour then place the mixture in the centre. Gather up the sides loosely and tie with string, leaving enough air in the parcel for the pudding to expand.
5. Very carefully, lower the pudding into the pan. Reduce the heat and simmer for 2½ to 3 hours. You will need to top up the water to stop it boiling dry.
6. Drain well and untie the cloth. Ease the pudding away from the cloth and put onto a large plate. Serve very hot.

BANANA FRITTERS

You will need:

50 g (2 oz) plain flour
¼ teaspoon salt
1 egg
110 ml (4 fl oz) milk
¼ teaspoon baking powder
2 large bananas
Lime or lemon juice

Vegetable oil for deep frying
Sugar for tossing

Method:

1. Sift the flour, baking powder and salt into a large mixing bowl. Make a well in the centre and break the egg into it. Add half the milk and gently stir to mix. Continue adding the milk a little at a time and stirring constantly to combine the flour, until all the milk is added and the batter is smooth and lump free. Continue to beat for 5 minutes then cover the bowl with a plate.
2. Peel and cut the bananas at a diagonal angle into 6 mm (¼ in) thick pieces. Coat thoroughly with batter.
3. Heat the oil to 180°C (350°F) in a pan or wok and fry the coated banana pieces for 1 to 2 minutes until golden brown. Drain on kitchen paper and sprinkle with lemon juice and sugar.

CHERRY COBBLER

This is an ideal dessert to make if you have a glut of cherries. That is, of course, if you can stop the birds getting them first. Tinned or preserved cherries work well too. Instead of using boiling water with the cornflour use the juice from the can or jar, warmed first.

You will need for the fruit: **Serves 6**

450 g (1 lb) fresh cherries
225 g (8 oz) sugar
1 tablespoon cornflour
225 ml (8 fl oz) boiling water

You will need for the cobbler:

125 g (5 oz) plain flour
1 teaspoon baking powder
110 g (4 oz) sugar
Pinch of salt
3 tablespoons melted butter
110 ml (4 fl oz) milk

Method:

1. Preheat the oven to 180°C (350°F, gas mark 4) and place cherries in the bottom of a 22 cm (9 in) square baking tin.
2. In a small bowl, mix together the sugar and cornflour then gradually stir in the boiling water. Mix well. Pour the cornflour mixture over the cherries.
3. Sift together the flour and baking powder into a large mixing bowl then add the sugar, butter, salt and milk and mix to a stiff dough. Spread the dough over the cherries.
4. Bake for 45 minutes and serve hot.

POOR MAN'S PUDDING

I don't know why this Canadian recipe has this name. It doesn't seem exactly to skimp on ingredients. The method is slightly unusual, but fun to try.

You will need: **Serves 4**

1 egg, beaten
110 g (4 oz) sugar
110 g (4 oz) plain flour
3 tablespoons softened butter
1½ teaspoons baking powder
Pinch of salt
80 ml (3 fl oz) milk

You will need for the sauce:

50 g (2 oz) chopped nuts
225 ml (8 fl oz) water
2 teaspoons softened butter
400 g (14 oz) brown sugar

Method:

1. Preheat the oven to 180°C (350°F, gas mark 4).
2. Place all the sauce ingredients into a saucepan, bring to the boil and let it simmer for 8 to 10 minutes.
3. Meanwhile, cream the butter and sugar in a large bowl. Add the egg, beating well, then add the milk and beat again.
4. Sift the flour, salt and baking powder and fold into the batter, mixing well until thoroughly blended.

5. Pour the batter into an ovenproof dish or loaf tin then pour the hot sauce over the top of the batter.
6. Bake for 40 minutes until cooked and serve hot.

HAZELNUT PANCAKES

You will need: **Serves 4**

2 eggs
270 ml (9 fl oz) milk
3 tablespoons runny honey
80 g (3 oz) chopped hazelnuts
80 g (3 oz) plain flour
1 teaspoon baking powder
1 teaspoon ground cinnamon
Pinch of salt
Vegetable oil for frying

Method:
1. In a large mixing bowl, beat together the eggs, milk and honey. Add the hazelnuts and the sifted flour, baking powder, cinnamon and salt. Beat well to form a smooth, lump-free batter.
2. Heat enough oil to coat a frying pan. Pour in a little of the batter, tilting the pan to cover the base evenly with a thin layer. Fry over a moderate heat, shaking the pan gently so it doesn't stick. Using a palette knife, turn the pancakes over and cook the other side until golden.
3. Remove to a plate, cover and keep warm while you cook the rest of the pancakes.
4. To serve, fold the pancakes into four and garnish with freshly sliced bananas, honey, cream or ice cream.

MILK JAM

This recipe from South America is best described as a reduced custard, made without eggs. It can be used to spread on biscuits, toast or as a cake filling. If you are brave enough to try it, you might need the tips on removing burnt-on milk from saucepans in the cleaning section!

You will need:

1.5l (2½ pt) milk
400 g (14 oz) sugar
1 tablespoon vanilla essence
1 teaspoon bicarbonate of soda

Method:

1. Place all the ingredients in a large saucepan and bring to simmering point, stirring all the time.
2. Continue to simmer for about 2 hours, stirring often, until a thick brownish paste is obtained. Cool in a bowl.

BISCUITS, CAKES AND COOKIES

I suppose this is the area I always associate with using bicarbonate of soda and baking powder. The cakes and biscuits produced tend to be of the 'bready' variety, rather than light sponge cakes, but the flavours are great. You can make cakes a healthier option by reducing the sugar content and by using oatmeal and/or wholemeal flour in place of white flour. The recipes below come from a range of cultural traditions. Using bicarbonate of soda and baking powder in cooking allows for variation in ingredients and accommodates various diets and regimes.

SCHOOL DINNER OAT COOKIES

This recipe takes me back to my first teaching job, when the school cook provided freshly baked biscuits to go with jelly or blancmange. You must try these, they're brilliant!

You will need: **Makes 20 biscuits**

125 g (4½ oz) porridge oats
125 g (4½ oz) butter or margarine
125 g (4½ oz) castor sugar
125 g (4½ oz) plain flour
¼ teaspoon bicarbonate of soda
1½ teaspoons baking powder
35 g (1¼ oz) golden syrup

Method:
1. Cream together the margarine and sugar.
2. Add the oats and remaining dry ingredients.
3. Mix with golden syrup.
4. Portion into small balls and place on greased tins, allowing room to spread.
5. Bake in a moderate oven 180°C (350°F, gas mark 4) for 15 to 20 minutes.

OATMEAL COOKIES

You will need: **Makes 30 cookies**

225 g (8 oz) butter or margarine
225 g (8 oz) brown sugar
60 g (2 oz) non-fat dry milk powder
1 large egg
1 teaspoon vanilla
1 tablespoon hot water
½ teaspoon bicarbonate of soda
Pinch of salt
225 g (8 oz) plain flour
350 g (12 oz) oatmeal

Optional extras:

100 g (3½ oz) raisins
100 g (3½ oz) walnut pieces
200 g (7 oz) chocolate chips

Method:

1. Preheat the oven to 180°C (350°F, gas mark 4).
2. Cream the butter or margarine with the sugar and milk powder, then add egg and vanilla.
3. Mix the water with baking soda and stir into the mix.
4. Fold in the salt, flour, oatmeal and raisins, walnuts or chocolate chips.
5. Roll dough into small balls and place on a greased baking sheet. Press flat.
6. Bake for 12 to 15 minutes, until golden brown.

GINGER BISCUITS

Syrup can get a bit messy and sticky, but this method saves the washing up.

You will need: **Makes about 24 biscuits**

150 g (5 oz) golden syrup
350 g (12 oz) plain flour
275 g (10 oz) sugar
110 g (4 oz) margarine
2 teaspoons ground ginger
1 teaspoon bicarbonate of soda

1 egg
Pinch of salt

Method:
1. Preheat the oven to 180°C (350°F, gas mark 4).
2. Put a small saucepan onto the scales and pour in the golden syrup.
3. Add the margarine and melt them together over a moderate heat.
4. Sift the flour, salt, ginger and bicarbonate of soda into a bowl. Beat in the egg and stir in the syrup mixture. Mix to a smooth dough.
5. Take teaspoons of the mix and roll into flattened balls. Place these on a baking sheet with room to spread.
6. Bake for about 15 minutes and cool them on the baking sheet. They can then be stored in an airtight tin quite successfully.

CHOCOLATE AND RAISIN COOKIES

You will need: **Makes about 24 cookies**

150 g (5½ oz) margarine
60 g (2 oz) soft light brown sugar
1 medium egg, beaten
80 g (3 oz) plain or wholemeal flour
1 teaspoon mixed spice
½ teaspoon bicarbonate of soda
150 g (5½ oz) rolled oats
40 g (1½ oz) raisins
40 g (1½ oz) chocolate chips
2 tablespoons water

Method:
1. Preheat the oven to 180°C (350°F, gas mark 4).
2. Cream together the margarine and sugar. Beat in the egg and water.
3. Sift together the flour, spice and bicarbonate of soda, tipping any bran in the sieve into the bowl and then stir in the oats, raisins and chocolate.
4. Add the egg mixture and fold in.
5. Line the baking sheets with greaseproof paper and place walnut-sized pieces of the mixture on the sheets,

with room to spread.
6. Bake for 10 to 15 minutes, or until golden.

CHOCOLATE, WALNUT AND DATE BROWNIES

This is a superior, adult version of a traditional recipe. I used 85% cocoa butter and chocolate and the cakes were delicious without being oversweet: an instant hit with the grown-up family. You can make one large cake and cut into squares when cool, or use paper cases for individual brownies.

You will need: **Makes about 16 brownies**

225 g (8 oz) plain chocolate or chocolate chips
80 g (3 oz) chopped dates with 6 tablespoons water
1 tablespoon baking powder
½ teaspoon bicarbonate of soda
60 g (2 oz) soft brown sugar
1 medium egg, beaten
175 g (6 oz) self-raising flour
3 tablespoons skimmed milk
40 g (1½ oz) pecan nuts or walnuts

Method:
1. Preheat the oven to 180°C (350°F, gas mark 4). Grease a baking tin.
2. Melt the chocolate over a pan of simmering water or in a microwave. Place the dates in a pan with the water, and simmer gently until they soften.
3. Sift the flour, baking powder and bicarbonate of soda into a bowl. Add the egg, milk, sugar and nuts.
4. Mix all the ingredients, including the chocolate and the date purée in a large bowl. This will be quite a sloppy batter.
5. Put small spoonfuls into cases or onto the prepared tin. Bake in the oven for 15 to 20 minutes, until just firm to the touch, before cooling on a wire tray.

COURGETTE CAKE

This cake has a wonderfully moist texture and seems like a good way to use up a glut of courgettes in the summer.

You will need:

225 g (8 oz) courgettes, grated
2 eggs
110 ml (4 fl oz) vegetable oil
The grated zest and juice of an orange
110 g (4 oz) caster sugar
225 g (8 oz) self-raising flour
40 g (1½ oz) cocoa powder
½ teaspoon bicarbonate of soda
½ teaspoon baking powder

To finish you will need:

200 g (7 oz) pack extra light cream cheese
1 tablespoon orange juice and zest of ½ orange
1 tablespoon icing sugar
Grated chocolate (optional)

Method:

1. Preheat the oven to 180°C (350°F, gas mark 4). Grease and line an 8 cm (3 in) deep loose-bottomed cake tin or loaf tin.
2. Put the courgettes through a sieve to squeeze out any excess liquid.
3. Beat the eggs with the oil, orange rind, juice and sugar in a large bowl.
4. Sift in the flour, cocoa, bicarbonate of soda and baking powder and beat to combine.
5. Fold in the courgettes and spoon the mix into the prepared tin. Bake for 40 minutes until risen and firm to touch.
6. When cool, beat together the cream cheese, orange and icing sugar and cover the top of the cake with this mixture. Sprinkle on the chocolate.

DATE AND GINGER CAKE

You will need: **Makes about 16 pieces**

225 g (8 oz) plain flour
1 teaspoon bicarbonate of soda
1 heaped teaspoon ground ginger
1 heaped tablespoon finely chopped stem ginger
1 egg
110 g (4 oz) margarine
110 g (4 oz) soft brown sugar
110 g (4 oz) black treacle
175 g (6 oz) chopped dates
150 ml (5 fl oz) water

Method:

1. Preheat the oven to 190°C (375°F, gas mark 5). Grease and line a small roasting tin or baking tray.
2. Put the margarine, sugar, treacle, water and dates into a large saucepan. Melt together and remove the pan from the heat.
3. When cooled slightly, stir in the sifted flour, bicarbonate of soda and ginger.
4. Beat in the egg. Mix well and pour into the prepared tin.
5. Bake for 35 minutes or until firm to the touch then leave to cool in the tin. Cut into 16 slices when cold.

CHOCOLATE SPONGE CAKE (EGGLESS)

Bicarbonate of soda can be used to replace eggs in a recipe, so if you have a reason not to eat eggs this might well be the answer. The basic recipe can be flavoured with other things, such as fresh lemon juice and zest or coffee and nuts. Since I can't eat a lot of cakes, I'm afraid chocolate always gets my vote.

You will need:

175 g (6 oz) self-raising flour
2 heaped teaspoons baking powder
25 g (1 oz) cocoa
80 g (3 oz) sugar
125 ml (4 fl oz) melted margarine
325 ml (11 fl oz) cold water

Method:

1. Preheat the oven to 190°C (375°F, gas mark 5).
2. Sift the flour, cocoa and baking powder together in a bowl.
3. Mix the sugar and melted margarine and add the water.
4. Fold in the dry ingredients and pour into two greased and lined tins.
5. Bake for about 25 minutes, or until a cocktail stick inserted in the centre comes out clean. Cool on a wire tray and decorate.

Optional filling and topping:
(Halve the quantity for filling only)

125 g (4½ oz) margarine
170 g (6 oz) icing sugar
Flavouring of choice, e.g. 25 g (1 oz) cocoa, mixed with
1 tablespoon very hot water
1 tablespoon instant coffee and some nuts

MUESLI CAKE (EGGLESS)

This is one of the healthier cakes you can make, being high in fibre and low in fat and sugar.

You will need:

175 g (6 oz) wholemeal or plain flour
175 g (6 oz) unsweetened muesli
125 g (4½ oz) brown sugar
175 g (6 oz) sultanas
2 tablespoons malt extract
3 teaspoons baking powder
250 ml (8 fl oz) apple juice
2 cooking apples, peeled, cored and grated

Method:

1. Put the muesli, sugar, malt extract, apple juice and sultanas in a large bowl and leave aside for at least 30 minutes.
2. Preheat the oven to 180°C (350°F, gas mark 4) and grease and line one 18 cm (7 in) cake tin while you wait.
3. Add the apple and flour to the soaked ingredients and

sift in the baking powder. Mix thoroughly.
4. Turn into the prepared tin and bake for 1 hour and 45 minutes to 2 hours, until a cocktail stick inserted into the centre comes out clean.
5. Cool on a wire tray. Decorate with walnuts or almonds for a special effect.

DORSET APPLE CAKE (EGGLESS)

This is another family favourite, adapted for those of us who must limit their egg intake. I usually serve it hot, accompanied by crème fraiche, cream or custard, but any leftovers are equally good served cold.

You will need:

3-4 eating apples
115 g (4 oz) margarine or butter
115 g (4 oz) brown sugar
115 g (4 oz) sultanas (optional)
2 level teaspoons bicarbonate of soda
1 tablespoon boiling water
225 g (8 oz) plain flour
½ teaspoon ground cinnamon

Method:
1. Preheat the oven to 180°C (350°F, gas mark 4).
2. Place the flour and cinnamon in a bowl with the sugar. Chop up the margarine or butter and add to the bowl. You don't need to rub it in.
3. Core and chop the apples into small chunks. Add these to the bowl with the raisins. This will look very lumpy, but that doesn't matter.
4. Mix the bicarbonate of soda with the water and add this to the mix. Put the mix into a cake tin and press down gently.
5. Bake for 45 minutes to 1 hour, until golden brown.

BUTTER RICE CAKE

This might appeal if you are fed up with, or can't eat, wheat flour.

You will need:

175 g (6 oz) rice cake flour
125 g (4½ oz) softened butter
1 teaspoon vanilla extract
175 g (6 oz) caster sugar
3 eggs, lightly whisked
½ teaspoon baking powder
¼ teaspoon bicarbonate of soda
1 teaspoon mixed spice
110 ml (4 fl oz) milk

Method:

1. Preheat the oven to 150°C (300°F, gas mark 2).
2. Sift together the rice cake flour, baking powder, bicarbonate of soda and mixed spice.
3. Place the butter, sugar, eggs and milk into a bowl and beat well.
4. Fold in the dry ingredients and beat until well combined.
5. Grease a 20 cm (8 in) round cake tin and line with baking paper.
6. Spoon in the mixture and bake for 55 minutes or until cooked. Allow to cool slightly, then turn onto a wire rack.

CARROT CAKE

This makes a large cake with up to 12 portions. If you prefer, you can use a reduced quantity and cook as individual muffins, as described later in this chapter.

You will need:

180 ml (6 fl oz) oil
3 eggs
225 g (8 oz) sugar
280 g (10 oz) plain flour
270 g (9½ oz) grated raw carrots
¾ teaspoon bicarbonate of soda
1½ teaspoons baking powder
2 teaspoons cinnamon
1 eating apple, peeled and grated or 25 g (1 oz) crushed, canned pineapple
60 g (2 oz) chopped walnuts or pecans
1 teaspoon salt

For the icing you will need:

110 g (4 oz) low fat cream cheese
1 tablespoon melted margarine or butter
3 tablespoons icing sugar

Method:

1. Preheat the oven to 180°C (350°F, gas mark 4).
2. Sift together the flour, salt, baking powder, bicarbonate of soda and cinnamon.
3. Stir in the sugar and nuts.
4. In a separate bowl, whisk the eggs with the oil.
5. Fold into the dry ingredients with the carrot and fruit. Don't overmix or you will end up with a rubbery texture. Just moisten the flour.
6. Put into a greased, rectangular baking pan for about 45 to 50 minutes (20 minutes for individual cakes).
7. Ice when cooled by combining the three ingredients and covering the top of the cake

BANANA CAKE

This can be iced or left plain.

You will need:

125 g (4½ oz) butter
200 g (7 oz) sugar
1 egg
175 g (6 oz) plain flour
2 tablespoons milk
1 teaspoon bicarbonate of soda
1 teaspoon baking powder
3 mashed ripe bananas
Pinch of salt

Method:

1. Preheat the oven to 180°C (350°F, gas mark 4) and prepare a 20 cm (8 in) square cake tin.
2. Cream the butter and sugar until light and then add the egg and milk.
3. Sift in the flour, bicarbonate of soda and baking powder.
4. Add the bananas and mix well.

5. Put the mixture into the cake tin and bake for 30 to 40 minutes.

PARKIN

This cake hails from the North of England. When we lived in Yorkshire the locals ate it with cheese and apple, as they do fruit cake.

You will need:

120 g (4½ oz) butter
120 g (4½ oz) medium oatmeal
225 g (8 oz) porridge oats
120 g (4½ oz) plain flour
3 teaspoons ground ginger
½ teaspoon bicarbonate of soda
¼ teaspoon baking powder
Pinch of salt
2 eggs
150 ml (¼ pint) treacle
15 ml (1 tablespoon) honey
75 ml (3 fl oz) milk
75 ml (3 fl oz) apple juice

Method:

1. Preheat the oven to 170°C (325°F, gas mark 3). Grease and line a square 20 cm (8 in) baking tin.
2. Mix the flour, oats, salt, ginger, baking powder and bicarbonate of soda together in a large bowl.
3. Put the butter, treacle and honey in a saucepan and bring to the boil, stirring to mix.
4. Add to the dry ingredients and mix.
5. Put the milk, eggs and apple juice into the pan and heat gently. It may curdle, but don't worry.
6. Add this mix to the bowl, stir well and pour into the baking tin.
7. Bake for about 30 to 35 minutes. Cool and cut into squares.

BARM BRACK

This fruit teabread is made without butter or margarine. You need to plan ahead and soak the fruit overnight.

You will need:

- 140 ml (5 fl oz) cold tea
- 225 g (8 oz) mixed fruit (raisins, sultanas, etc.)
- 110 g (4 oz) demerera sugar
- 50 g (2 oz) walnut pieces
- 50 g (2 oz) glacé cherries, sliced
- 1 small egg
- 1 tablespoon milk
- 225 g (8 oz) wholemeal flour (or half wholemeal, half white)
- 2 teaspoons baking powder

Method:

1. Soak the mixed fruit in the tea and sugar overnight.
2. Preheat the oven to 170°C (325°F, gas mark 3).
3. Gradually mix in the egg, milk, cherries and nuts to the tea-soaked fruit. Add the flour and baking powder and turn into a prepared loaf tin.
4. Bake for one hour or so, until cooked.

GINGERBREAD

This old family favourite can be jazzed up with added fruit or nuts. It is easy to make and keeps fairly well. It is also good as a pudding, as my mother would have it, with custard or a drop of brandy.

You will need:

- 225 g (8 oz) plain flour
- 1 teaspoon bicarbonate of soda
- 2 teaspoons ground ginger
- 150 g (5 oz) golden syrup
- 110 g (4 oz) margarine
- 50 g (2 oz) brown sugar
- 2 eggs
- 2 tablespoons milk
- 110 g (4 oz) sultanas (optional)

Method:

1. Preheat the oven to 170°C (325°F, gas mark 3) and grease and line an 18 cm (7 in) square tin.
2. Sift the flour, bicarbonate of soda and ginger together in a bowl.
3. Warm the golden syrup in a saucepan with the margarine and brown sugar until the ingredients have melted. Add the milk and leave to cool.
4. Beat the eggs into the mixture in the saucepan and then mix this with the dry ingredients to form a thick batter. Stir in the sultanas.
5. Pour into the lined tin and bake for 1 to 1¼ hours, or until cooked through.

YOGHURT CAKE

This recipe comes from Turkey and uses semolina in an interesting way. The syrup is very sweet, so you may wish to cut down the amount used. I think warmed honey would be a good substitute.

You will need:

175 g (6 oz) semolina
80 g (3 oz) plain flour
250 g (9 oz) plain yoghurt (low fat for preference)
200 g (7 oz) sugar
1 egg
1 teaspoon bicarbonate of soda
1 teaspoon ground cinnamon

You will need for the syrup:

225 g (8 oz) sugar
225 ml (8 fl oz) water

or

110 ml (4 fl oz) warmed honey

Method:

1. Preheat the oven to 180°C (350°F, gas mark 4) and prepare a large, square ovenproof dish or tin.
2. Blend together the yoghurt and sugar and then beat in the egg.
3. Add the semolina, flour, bicarbonate of soda and

cinnamon. Beat until well mixed.
4. Pour the mixture into the prepared tin and bake for about 45 minutes or until golden brown.
5. Just before the end of the cooking time, put the sugar and water into a saucepan, bring to the boil and simmer for a few minutes (skip this if using honey).
6. When you remove the cake from the oven, spike it all over with a fork. Pour the syrup (or warm honey) over the warm cake. Allow the cake to cool and absorb most of the sweetener. Cut into squares and serve cold.

YAOURTOPITA

This Greek cake uses yoghurt, as the name may suggest. The syrup is very sweet, so you may wish to tone it down a little, or use an alternative, such as orange juice.

You will need:

125 g (4½ oz) plain flour
½ teaspoon bicarbonate of soda
1 teaspoon baking powder
175 g (6 oz) sugar
110 g (4 oz) butter
3 eggs, separated
120 ml (4 fl oz) plain yoghurt

For the syrup you will need:

225 g (8 oz) sugar
120 ml (4 fl oz) water
Juice and zest of ½ a lemon

Method:

1. Preheat the oven to 170°C (325°F, gas mark 3) and grease and line a 20 cm (8 in) cake tin.
2. Sift the flour with the bicarbonate of soda and baking powder into a bowl.
3. In a separate bowl, cream the butter with the sugar until light and fluffy. Beat in the egg yolks, one at a time and then the yoghurt.
4. In another bowl whisk the egg whites until stiff.
5. Add the dry ingredients by sifting again into the egg yolk mix and then fold in the egg whites.

6. Pour into the prepared tin and bake for about 45 minutes.
7. Turn out onto a wire tray and allow to cool.
8. Boil the sugar, water and lemon juice for about 5 minutes and pour over the cooled cake.

CHOCOLATE MALT CAKE

You will need:

175 g (6 oz) wholemeal flour
2 teaspoons baking powder
25 g (1 oz) cocoa
175 g (6 oz) molasses or brown sugar
2 eggs, beaten
120 ml (4 fl oz) oil
120 ml (4 fl oz) milk
2 tablespoons malt extract

For the topping and filling you will need:

125 g (4½ oz) low fat cream cheese
75 g (3 oz) plain chocolate

Method:

1. Preheat the oven to 170°C (325°F, gas mark 3) and grease and line two sandwich tins (18 to 20 cm/7 or 8 in).
2. Put the sugar, oil, eggs, milk and malt extract into a large bowl and mix well.
3. Add the flour and sift in the cocoa and baking powder. Beat until smooth.
4. Divide between the sandwich tins and smooth flat.
5. Bake for 25 to 30 minutes, until springy.
6. Turn out and cool on a wire rack.
7. Melt the chocolate over a bowl of hot water or in a microwave. Beat the cheese until smooth and add the melted chocolate. Beat well.
8. Put half the filling between the two parts of the cake and spread the rest on the top. Make a pattern with a fork to finish.

BANANA NUT SLICES

This cake uses a more traditional format and is best made with an electric whisk. You could use a food processor.

You will need:

125 g (4½ oz) plain flour, or wholemeal
125 g (4½ oz) brown sugar
125 g (4½ oz) margarine or butter
2 eggs
2 teaspoons baking powder
2 mashed bananas
125 g (4½ oz) chopped hazelnuts

Method:

1. Preheat the oven to 190°C (375°F, gas mark 5). Grease and line a 20 cm (8 in) square tin.
2. Cream the fat with the sugar until light and fluffy. Beat in the eggs, one at a time.
3. Sift in the flour and baking powder and fold into the mix with the banana and nuts.
4. Spread evenly across the baking tin and cook for 20 minutes, or until springy.
5. Leave to cool slightly in the tin and then cut into 16 slices, before cooling on a wire rack.

CHOCOLATE FUDGE CAKE

We can't be health conscious all the time. Even people with diabetes are allowed a birthday cake (well, a small piece, anyway). This delicious recipe fits the bill perfectly.

You will need:

250 g (8 oz) plain flour
125 g (4 oz) soft brown sugar
125 g (4 oz) plain chocolate
125 g (4 oz) margarine
125 g (4 oz) caster sugar
2 eggs, separated
300 ml (½ pint) semi-skimmed milk
1 teaspoon bicarbonate of soda

For the chocolate fudge topping you will need:

125 g (4 oz) icing sugar
1 tablespoon milk
25 g (1 oz) margarine
1 tablespoon cocoa

Method:

1. Preheat the oven to 180°C (350°F, gas mark 4) and grease and line a 20 cm (8 in) baking tin.
2. Put the chocolate, some of the milk and the brown sugar into a saucepan and melt slowly. Stir in the rest of the milk.
3. Whisk the margarine with the caster sugar until light and fluffy and then beat in the egg yolks, one at a time.
4. Sift the flour and bicarbonate of soda together into the creamed ingredients and add the chocolate and milk. Beat until smooth.
5. Whisk the egg whites until stiff and then fold carefully into the batter.
6. Bake for about 45 minutes, until springy.
7. Turn out and cool slightly on a wire rack while you make the icing.
8. Put the fat and the milk into a saucepan and heat gently until melted.
9. Sift in the icing sugar and cocoa and mix well. Spread over the cake while still slightly warm.

CHERRY AND ALMOND SPONGE

You will need:

175 g (6 oz) plain flour
175 g (6 oz) caster sugar
175 g (6 oz) margarine
3 eggs
1 teaspoon baking powder
80 g (3 oz) ground almonds
110 g (4 oz) glacé cherries, cut into quarters
1 teaspoon almond essence
1 tablespoon milk

Method:

1. Preheat the oven to 180°C (350°F, gas mark 4). Grease and prepare a loaf tin or fluted cake mould (savarin tin).
2. Cream the margarine and sugar until fluffy, then beat in the eggs, one at a time.
3. Add the chopped cherries, milk, almonds and almond essence.
4. Sift in the flour and baking powder and fold into the mix.
5. Spoon into the prepared tin and bake in the centre of the oven for 25 to 30 minutes, or until cooked.
6. Cool on a wire tray.

CIDER CAKE

This is another recipe to prepare the night before you want to bake.

You will need:

350 g (12 oz) mixed dried fruit (raisins, currants, sultanas)
750 ml (1¼ pt) cider
225 g (8 oz) plain flour
1 teaspoon baking powder
½ teaspoon mixed spice
110 g (4 oz) butter or margarine
110 g (4 oz) soft dark brown sugar
50 g (2 oz) mixed peel
Finely grated zest of one orange
2 eggs, beaten
Sugar for topping
Pinch of salt

Method:

1. Soak the dried fruit overnight in the cider in a large bowl.
2. Preheat the oven to 170°C (325°F, gas mark 3). Grease and base line a 23 cm (9 in) round deep cake tin.
3. Sift together the flour, baking powder, pinch of salt and mixed spice into a bowl. Mix in the butter or margarine into a crumb-like texture. Stir in the sugar, mixed peel and grated orange rind.
4. Mix all these ingredients thoroughly and then make a hollow in the centre of the mix. Beat in the eggs and the fruit and cider.

5. Mix until it reaches a soft consistency. Pour into the cake tin, levelling off the top. Sprinkle a little sugar on top.
6. Bake for one hour, then reduce the oven temperature to 150°C (300°F, gas mark 2) and cook for a further hour or until the top of the cake is springy and the cake is shrinking slightly. If you are unsure, test with a skewer, which should come out clean.
7. Turn the cake out onto a wire rack to cool and then store in an airtight tin to let it mature for a few days.

BLUEBERRY POUND CAKE

You will need: **Makes 1 x 25 cm (10 in) cake**

225 g (8 oz) granulated sugar
80 ml (3 fl oz) oil
2 eggs
175 g (6 oz) plain flour
110 g (4 oz) fresh or frozen blueberries
½ teaspoon baking powder
¼ teaspoon bicarbonate of soda
Pinch of salt
110 ml (4 fl oz) sour cream
½ teaspoon lemon juice

You will need for the icing:

25 g (1 oz) icing sugar
2 teaspoons lemon juice

Method:
1. Preheat the oven to 180°C (350°F, gas mark 4) and lightly grease a 25 cm (10 inch) cake tin.
2. Place the sugar and oil in a large mixing bowl and mix until well blended. Add the eggs, one at a time, beating well after each addition.
3. Place the blueberries in a small bowl, add 2 tablespoons of the flour and toss well.
4. Sift the remaining flour, baking powder, bicarbonate of soda and salt in a large mixing bowl. Add this to the sugar mixture a little at a time, alternating with the cream.
5. Fold in the blueberry mixture and lemon juice then pour into the prepared tin.

6. Bake for 1 hour and 10 minutes or until a skewer inserted in the centre comes out clean. Remove from the oven and allow to cool in the tin for 10 minutes.
7. Mix together the icing sugar and lemon juice in a small bowl. Remove the cake from the tin, transfer to a plate and drizzle the icing sugar mixture over the top of the warm cake.

TAHITIAN SWEET BREAD

This exotic fruity cake uses dried fruit. If you use fresh fruit I suggest you reduce the orange juice.

You will need: **Makes 1 loaf**

175 g (6 oz) plain flour
110 g (4 oz) whole wheat flour
175 g (6 oz) sugar
1 teaspoon bicarbonate of soda
1 teaspoon baking powder
½ teaspoon salt
225 ml (8 fl oz) plain, low fat yoghurt
2 egg whites, lightly beaten
2 tablespoons vegetable oil
½ teaspoon crystallised ginger
Juice and zest of 2 oranges
110 g (4 oz) dried mango
110 g (4 oz) dried pineapple

Method:
1. Preheat the oven to 180°C (350°F, gas mark 4) and lightly grease a loaf tin.
2. In a large bowl, mix together the flours, sugar, bicarbonate of soda, baking powder and salt.
3. In a separate bowl, mix together the yoghurt, egg whites, oil, orange juice and ginger.
4. Add the yoghurt mixture to the flour mixture and stir until combined. Do not over-mix the batter.
5. Stir in the orange zest, mango and pineapple until just blended then pour into the prepared tin.
6. Bake for about 50 minutes or until a skewer inserted in the centre comes out clean. Allow to cool in the tin for 10 minutes before removing and cooling on a wire rack.

MUFFINS

I've always been confused by this word, as the muffins I remember from my youth were yeast-based, flat, pancake-like items that you toasted and smothered with butter. Nowadays most people seem to accept that a muffin looks like an overgrown fairy cake. Now I know why.
The English muffin is, indeed, a thick, flat bun-like product made with yeast, which looks like a crumpet or a pikelet. The American muffin is leavened with bicarbonate of soda, or baking powder, or both, and may contain all manner of ingredients, including flour, sugar, some sort of fat or oil and milk. A variety of other flavourings can be added. These muffins became popular after the introduction of baking powder in America during the late 19th century. One of the drawbacks in the early days was that they grew stale very quickly, so commercially they were not as viable as other cake products, but modern preservatives have resulted in a longer shelf-life and the now, seemingly ubiquitous muffin outlet.

Compared with the calorific monsters which include doughnuts and Danish pastries, low fat muffins seem like good news. Of course, to stay fresh they need to contain more fat and sugar, which somewhat spoils the illusion of healthy eating, and with all the possible additions and combinations, such as chocolate chip, coconut and banana, the yoghurt or skimmed milk isn't going to count for much. Being confined to a low fat, low sugar, high fibre diet myself, I'm not a connoisseur, but I'm amazed how quickly my daughter can devour one of the huge, chocolate muffins you buy in coffee shops. They always stick to the roof of my mouth, which is just nature's way of telling me not to eat them.

Muffins are not the easiest of cakes to cook if you are a beginner. The technique of mixing a batter is quite different from that of creaming fat and sugar together, leavening with beaten egg and adding flour and flavourings. Home-made muffins are a lot more interesting, however, than the commercial ones and you can at least control what goes into them.

Some recipes contain both baking powder and bicarbonate of soda. In this case it is the baking powder that does most of the leavening. The bicarbonate of soda neutralises the acids in the recipe and adds some leavening. When you use either product, make sure you sift or whisk them with the other dry ingredients before you add them to the batter, to ensure uniformity. Too much baking powder can cause the batter to taste bitter and can also cause rapid rising, followed by collapse. This happens when the air bubbles in the batter grow too large and then pop, causing the batter to fall. Cakes like this will have a coarse, crumby structure with a stodgy middle. Too little baking powder will give you a rubbery cake which doesn't rise to the occasion.

MUFFIN PERFECTION
The perfect muffin is symmetrical with a domed top. The surface is bumpy and the volume of the batter should almost double during baking, after which it should feel light for its size.

SWEET OR SAVOURY
I have always associated muffins with sweet rather than savoury ingredients, but you will find recipes for both below. They are traditionally served warm for breakfast, when ingredients such as oats, bran, yoghurt or fruit make for a different way of eating cereal. As an alternative, savoury muffins can go well with salad or soups for lunch.

The reason why some muffins rise more than others, apart from the baking powder/bicarbonate of soda quantities used, is that less sugar and butter makes a bread-like muffin, whereas a higher sugar and fat content creates a cake-like muffin.

Bread-like muffins are made using the two bowl method. This batter can be assembled and baked quickly, usually in about 20 minutes. One bowl is used to mix all the dry ingredients together while the second contains all the wet ingredients. The fat used with these muffins is usually in liquid form, either oil or melted butter. When the wet and dry ingredients have been prepared separately, they are combined quickly, with little mixing, because too much overdevelops the gluten in the flour. There is, however, a tendency to overmix because the ratio of liquid to flour is quite high. Only 15 seconds are needed to moisten the ingredients. The batter should still be lumpy, with traces of flour. These will disappear as the batter continues to blend as it bakes.

PAPER CASES

Using paper muffin cup liners is the best idea for cooking and presentation. These not only make cleaning up easier, they also help to keep the muffins moist and prevent them from drying out.

BAKING

Muffins should be baked in the centre to top of a preheated oven and are cooked when a toothpick inserted in the middle comes out clean and the edges start to come away from the sides of the pan, usually after 20 to 25 minutes at 180° to 200°C (350° to 400°F, gas mark 4 to 6). Spoon the muffin batter into the paper using two spoons or, better still, an ice cream scoop, only filling each one to just over halfway. Handle the batter as little as possible, otherwise you will end up with tough muffins.

MUFFIN TROUBLESHOOTING

PROBLEM	POSSIBLE REASON
Muffins have tunnels inside and are rather dry.	• The batter was overmixed. • They were overbaked and/or the oven was too hot. • There is too much flour and/or too little liquid.
Muffins have an uneven shape.	• There is too much batter in each cup (fill only ½ to ⅔ full). • Oven temperature was too high.
Tops are brown but the muffin is not cooked through.	•The oven temperature was too high. • The oven rack was not in the centre of the oven.
Muffin does not rise sufficiently.	• The oven temperature was too low. • The batter was overmixed or an incorrect amount of leavening was used.

The following recipes contain a variety of muffins to try. Some are healthier than others. Quantities of ingredients such as flour, fat and sugar may vary, giving different textures for bread-like or cake varieties. The American muffin recipes tend to have higher sugar content, so take care if you are dieting. Low fat does not always mean low sugar as well! It is best to make small amounts that can be eaten immediately, rather than batches, although you can freeze them. Fresh (and still warm) is best, in my opinion.

As always, use either the metric measures or the imperial, but don't mix the two, as there are always discrepancies between them. Be aware also that ovens vary, so go by your oven guide for approximate cooking times. Fan ovens often cook faster, so I'm told. My Aga has a will of its own and no instant temperature control and I can't judge times for different types of ovens purely from my own experience. Expect to experiment, rather than to produce the perfect muffins first time.

MUFFINS MADE WITH BUTTER

A BASIC BUTTER MUFFIN RECIPE

This should make 6 large muffins. If you make small ones, adjust the cooking time by about 8 to 10 minutes.

You will need:

150 g (5½ oz) plain flour
40 g (1½ oz) caster sugar
1 egg
50 g (2 oz) melted butter
110 ml (4 fl oz) milk
½ teaspoon baking powder
Pinch salt
½ teaspoon vanilla extract (optional)

Suggested additions:

110 g (4 oz) blueberries and 50 g (2 oz) pecans, chopped finely
75 g (3 oz) glacé cherries
50 g (2 oz) chopped dates (sprinkle with flour to stop them sticking together)
75 g (3 oz) raisins
1 tablespoon strong instant coffee and 75 g (3 oz) walnuts

Method:

1. Preheat the oven to 200°C (400°F, gas mark 6).
2. Sift together the flour, baking powder and salt into a bowl.
3. Mix the egg, sugar, milk, vanilla and the cooled, melted butter in a second bowl.
4. Sift the flour mix (a second time) into the egg mix. Using a large spoon, fold the wet ingredients into the dry ones. This should take no more than 15 seconds and the mixture may appear lumpy.
5. Fold in any fruit or nuts very briefly, then spoon balls of the mixture into six paper baking cases.
6. Place on a baking tray or in a bun tin on a high shelf for about 25 minutes. Test whether the inside is cooked with a cocktail stick, which should be dry when removed.
7. Cool on a wire tray. If you are going to ice them, let them cool first.

CHOCOLATE CHIP MUFFINS

As you can see, these have a lot more sugar.

You will need:

240 g (8 oz) plain flour
40 g (1½ oz) light-brown sugar
40 g (1½ oz) white sugar
1 teaspoon baking powder
Pinch salt
80 ml (3 fl oz) milk
50 g (2 oz) butter, melted and cooled
1 egg, lightly beaten
½ teaspoon vanilla essence (optional)
25 g (1 oz) milk chocolate chips
50 g (2 oz) walnuts or pecans, chopped

Method:

1. Preheat the oven to 200°C (400°F, gas mark 6).
2. In a large bowl, sift together the flour, baking powder, salt and add sugars.
3. In another bowl, stir together the milk, eggs, butter and vanilla.
4. Combine the two mixes, folding with a spoon.
5. Stir in the chocolate chips and nuts.
6. Spoon the batter into the paper cases and bake for about 20 minutes or until a cocktail stick comes out clean.
7. Remove the muffins to a wire rack.

BANANA MUFFINS

You will need:

175 g (6 oz) plain flour
½ teaspoon bicarbonate of soda
½ teaspoon baking powder
Pinch salt
1 to 2 ripe bananas (depending on size)
80 g (3 oz) sugar
1 small egg, lightly beaten
40 g (1½ oz) butter, melted

You will need for topping:

40 g (1½ oz) brown sugar
Pinch ground cinnamon
2 teaspoons melted butter
25 g (1 oz) chopped nuts (optional)

Method:

1. Preheat the oven to 190°C (375°F, gas mark 5).
2. Sift and mix the dry ingredients.
3. Mash the bananas and mix with the sugar, egg and butter. Stir into the dry mixture until just combined.
4. Put into muffin cases with two spoons and bake for about 20 minutes, until cooked.
5. Combine the topping ingredients and put on the muffins while still hot.

BLUEBERRY AND ORANGE MUFFINS

You will need:

1 orange
60 ml (2 fl oz) orange juice
1 egg
40 g (1½ oz) butter
150 g (5½ oz) plain flour
60 g (2 oz) natural bran
60 g (2 oz) white sugar
½ teaspoon baking powder
½ teaspoon bicarbonate of soda
60 g (2 oz) blueberries

Method:

1. Preheat the oven to 190°C (375°F, gas mark 5).
2. Grate the zest of the orange and put the orange, with the seeds removed, into a blender with the orange juice. Blend to a purée.
3. In a separate bowl, mix the sifted flour, bicarbonate of soda and baking powder with the bran and sugar.
4. Combine the orange purée with the zest, egg and butter.
5. Mix the dry ingredients with the wet ones and then add the blueberries. Don't mix too much.
6. Spoon into paper cases. Bake for about 20 minutes or until brown.

BANANA AND NUT MUFFINS

This recipe uses a different method, more like a traditional sponge cake method, so the air is trapped by beating in the egg as well as by the raising agents in the ingredients.

You will need:

60 g (2 oz) butter or margarine at room temperature
120 g (4 oz) granulated sugar
1 large egg
1 large ripe banana, mashed
½ teaspoon vanilla essence (optional)
240 g (8 oz) plain flour
½ teaspoon salt
½ teaspoon baking powder
¼ teaspoon bicarbonate of soda
120 ml (4 fl oz) buttermilk (or milk with ½ tablespoon lemon juice)
60 g (2 oz) chopped pecans or walnuts

Method:

1. Preheat the oven to 200°C (400°F, gas mark 6).
2. Cream the butter or margarine and sugar until light and fluffy. Beat in the egg, a little at a time.
3. Add the banana and vanilla and beat until smooth.
4. Sift together the flour, salt, baking powder and bicarbonate of soda.
5. Fold the flour mixture into the batter, alternating with the buttermilk. Stir just until the dry ingredients are moistened and add the nuts.
6. Carefully spoon the batter into paper cases, filling about two-thirds full.
7. Bake for about 18 to 20 minutes, or until tops are lightly browned. Serve warm.

ONION AND CORNMEAL MUFFINS

These are savoury muffins, despite the sugar.

You will need:

150 g (5½ oz) yellow cornmeal
80 g (3 oz) all-purpose flour
25 g (1 oz) sugar
1½ teaspoons baking powder
½ teaspoon salt
120 ml (4 fl oz) milk
25 g (1 oz) butter, melted
60 g (2 oz) chopped spring onions
1 egg

Method:

1. Preheat the oven to 200°C (400°F, gas mark 6).
2. Sift together the flour, baking powder and salt in a large bowl and add the cornmeal and sugar.
3. In a separate bowl, beat the egg until frothy. Mix in the melted butter and milk.
4. Fold the two mixtures together quickly and mix in the onions to make a lumpy batter.
5. Put into paper cases and cook for 20 to 25 minutes.

MUFFIN RECIPES WITHOUT BUTTER

BANANA-CHOCOLATE CHIP MUFFINS

A very low-fat recipe, but high on the sugar content. The bananas make up for the lack of liquid.

You will need: **Makes 6 muffins**

2 medium, very ripe bananas
1 egg
50 ml (2 fl oz) low-fat buttermilk (or milk with ½ tablespoon lemon juice)
50 g (2 oz) granulated sugar
50 g (2 oz) brown sugar
150 g (5 oz) plain flour
½ teaspoon bicarbonate of soda
½ teaspoon salt
20 g (¾ oz) chocolate chips
15 g (½ oz) chopped walnuts (optional)

Method:

1. Preheat the oven to 180°C (350°F, gas mark 4).
2. In a large bowl, mash the bananas with a fork. Whisk in the egg, milk, granulated sugar and brown sugar.
3. In a separate bowl, sift together the flour, bicarbonate of soda and salt.
4. Add the flour to the wet ingredients. Fold quickly with a spoon.
5. Fold in the chocolate chips and walnuts.
6. Spoon equal amounts of batter into paper cases. Bake for 30 minutes or until brown and cooked.
7. Remove to a wire rack to cool.

FRUITY BREAKFAST MUFFINS

These muffins are packed with fibre but are low in calories.

You will need:

80 g (3 oz) wholemeal flour
160 g (6 oz) plain flour
Pinch of cinnamon
Pinch of salt
2 teaspoons bicarbonate of soda
60 g (2 oz) brown sugar
2 tablespoons wheatgerm
3 tablespoons orange juice
Grated zest of ½ an orange
60 ml (2 fl oz) oil
1 egg
160 g (6 oz) raisins
240 ml (8 fl oz) plain, low fat yoghurt

Method:

1. Preheat the oven to 200°C (400°F, gas mark 6).
2. Sift together the flour, salt, bicarbonate of soda and cinnamon and then add the sugar and wheatgerm.
3. In a separate bowl, whisk the egg and add the oil, yoghurt, orange juice and zest.
4. Fold the two mixes together, just enough to combine roughly and add the raisins.
5. Spoon into paper cups and bake for about 20 minutes.

LEMON, YOGHURT AND OATMEAL MUFFINS

I can recommend these as tasty, healthy and easy to make. They are perfect for a high fibre, low fat diet. They got the thumbs up in our house, even without the syrup topping.

You will need:

175 g (6 oz) plain flour
120 g (4 oz) rolled oats
50 g (2 oz) granulated sugar
1 teaspoon baking powder
½ teaspoon bicarbonate of soda
Pinch salt
150 ml (5 fl oz) plain, low fat yoghurt
1 small egg
2 tablespoons lemon juice
Grated rind of ½ lemon

You will need for the syrup (optional):

30 g (1 oz) granulated sugar
½ fresh lemon rind, grated
1 tablespoon lemon juice

Method:

1. Preheat the oven to 200°C (400°F, gas mark 6).
2. In a large bowl put the rolled oats, granulated sugar, and sift in the flour, baking powder, bicarbonate of soda and salt.
3. In a separate bowl, mix together the yoghurt, egg, lemon juice and rind.
4. Add the wet ingredients to the dry, mixing only until roughly blended.
5. Divide between about 10 paper cases. The mix will look very dry, but this doesn't matter.
6. Bake for 20 minutes, or until cooked.
7. Prepare the syrup by putting the ingredients into a small saucepan. Bring to the boil and leave aside.
8. When the muffins are baked, prick all over with a toothpick and brush the warm syrup on the hot muffins.

PUMPKIN MUFFINS

These are perfect for Halloween, or for encouraging everyone to eat more pumpkin. You could use some of the flesh that you hollow out for the lantern. For added interest these can be iced or topped with a cream cheese frosting.

You will need:

175 g (6 oz) plain flour
1 teaspoon bicarbonate of soda
120 g (4 oz) sugar
½ teaspoon salt
½ teaspoon cinnamon
½ teaspoon nutmeg
Pinch ground ginger
120 ml (4 fl oz) oil
2 eggs
175 g (6 oz) puréed pumpkin

Method:

1. Preheat the oven to 180°C (350°F, gas mark 4).
2. Sift the flour, salt and bicarbonate of soda. Mix with the spices.
3. Lightly beat the eggs and add to the pumpkin, oil and sugar.
4. Combine the wet and dry ingredients without overmixing.
5. Spoon into paper cases and bake for about 25 minutes, or until a cocktail stick inserted into the centre comes out clean.

CHOCOLATE MUFFINS

Bicarbonate of soda has the effect of reddening cakes that have cocoa as an ingredient, giving rise to the name of Devil's food cake. These very chocolaty muffins are a deep, rich colour and flavour and also low in sugar. I forgot to put any in the first time I made them, and they still tasted okay, but they were even better with a little sugar.

You will need:

240 g (8 oz) plain flour
80 g (3 oz) cocoa
60 g (2 oz) sugar
1 teaspoon bicarbonate of soda
Pinch of salt
160 ml (6 fl oz) milk
40 ml (1½ fl oz) vegetable oil
1 egg
120 g (4 oz) chocolate chips

Method:

1. Preheat the oven to 200°C (400°F, gas mark 6).
2. Sift the flour, salt, cocoa and bicarbonate of soda together and add the sugar.
3. Mix together the egg, milk and oil in a separate bowl.
4. Spoon the two mixtures together, with half of the chocolate chips. Keep mixing to the minimum.
5. Spoon into paper cases and sprinkle the rest of the chocolate chips on the tops.
6. Bake for about 20 minutes.

CHILLI CORN MUFFINS

Be careful when preparing the chilli in this recipe. If you haven't used chillies before, be sure to keep your hands away from your face when handling, or you could end up in pain.

These are a savoury variety and could be served with soup as a change from bread.

You will need:

120 g (4 oz) plain flour
80 g (3 oz) cornmeal
25 g (1 oz) granulated sugar
1½ teaspoons baking powder
Pinch of salt
1 small egg
2 tablespoons cooking oil
100 ml (3½ fl oz) milk
1 small chilli, seeded and chopped finely

Method:

1. Preheat the oven to 200°C (400°F, gas mark 6).
2. Mix the flour, cornmeal, sugar, baking powder and salt in a bowl.
3. In a second bowl, mix the remaining ingredients, except the chilli.
4. Add the milk mixture to the dry ingredients and stir until just blended. Add the chilli and spoon the mix into paper cases.
5. Bake for 15 to 20 minutes, or until the top springs back when lightly pressed.

BREAD, SCONES AND PANCAKES

SCOTCH PANCAKES

You will need:

225 g (8 oz) plain flour
2 eggs
60 g (2 oz) caster sugar
120 ml (4½ fl oz) milk
1 teaspoon cream of tartar
½ teaspoon bicarbonate of soda
Pinch of salt

Method:

1. Sift the flour, cream of tartar, salt and bicarbonate of soda together into a large bowl and add the sugar.
2. Make a well in the centre and beat in the eggs and enough milk to make a thick batter.
3. Drop dessertspoons of batter onto a greased, preheated griddle or pan.
4. Cook until the tops of the pancakes are covered with tiny bubbles, then flip them over to cook the undersides until golden brown, about 4 minutes each side.
5. Cool between folds of a clean tea towel to keep soft while you cook the remaining batter. Serve with butter or choice of topping (see below for a suggestion).

PANCAKES WITH MAPLE APPLES AND PECAN

Make the Scotch pancakes, as above.

You will need for topping:

3 eating apples, peeled, cored and finely chopped
25 g (1 oz) butter
80 g (3 oz) pecan pieces
175 ml (6 fl oz) maple syrup

Method:

1. Place the butter in a saucepan and when melted, add the apple. Simmer for a few minutes.
2. Add the syrup and pecans. Bring almost to the boil, stirring occasionally.
3. Allow to cool.
4. To serve, place a spoonful of the apple mixture on a

pancake, top with a second pancake and spoon over more of the topping.

NAAN BREAD

Naan bread can be cooked with or without eggs. If you leave out the egg, substitute with 110 ml (4 fl oz) of extra yoghurt and ¼ teaspoon of bicarbonate of soda. You need a very hot oven to try to replicate the tandoor, or hot clay oven, used in India. If you have a separate grill you can use that as well.

You will need: **Makes 4**

1 teaspoon dried yeast
80 ml (3 fl oz) tepid milk
1 teaspoon sugar
225 g (8 oz) plain flour
1 teaspoon baking powder
¼ teaspoon salt
60 ml (2 fl oz) plain yoghurt
1 egg, beaten
2 tablespoons melted butter or oil

Method:

1. Mix together the yeast, warm milk and sugar and leave for about 15 minutes, until frothy.
2. Sift the flour, baking powder and salt into a large mixing bowl, make a well in the centre then pour in the yoghurt, egg (if using), half the melted butter or oil and the frothy yeast mixture.
3. Mix all the ingredients together well and knead thoroughly for about 10 minutes on a floured surface. Form into a ball.
4. Place in a large bowl. Cover with oiled cling film and put in a warm place to double in size.
5. Preheat the oven to 200°C (400°F, gas mark 6) and lightly grease and heat two baking trays. Preheat your grill as well.
6. Turn the dough onto a floured surface and divide into four equal pieces.
7. Leave the remainder covered in the bowl until you are ready to work it. Lightly knead each piece and roll out into

tear shapes, approximately 20 cm (8 in) by 12 cm (5 in).
8. Place on the greased trays and bake for 10 to 15 minutes. Alternatively, if you have a grill as well, bake for 4 to 5 minutes and then transfer to the grill to brown on each side for a minute. Serve hot.

CRANBERRY AND HONEY OAT BREAD

You will need: **Makes 1 loaf**

2 tablespoons honey
1 egg
2 tablespoons milk
1 teaspoon melted butter or margarine
350 g (12 oz) flour
150 g (5 oz) rolled (porridge) oats
1 dessertspoon baking powder
½ teaspoon salt
½ teaspoon ground cinnamon
100 g (4 oz) fresh or frozen cranberries
50 g (2 oz) chopped nuts (optional)

Method:
1. Preheat the oven to 180°C (350°F, gas mark 4) and grease a loaf tin.
2. Put the flour, oats, baking powder and salt in a bowl.
3. Beat together the egg, honey and milk and add to the flour and oats, stirring well. Fold in the cranberries and nuts.
4. Pour into the loaf tin and bake for about 75 minutes, or until crusty and hollow when knocked. Pour the melted butter over the hot loaf and turn out onto a cooling rack.

CHEESE BREAD

As with the Soda Bread recipe, milk and lemon juice may be substituted for buttermilk.

You will need:

200 g (7 oz) plain flour
250 g (9 oz) wholemeal flour
2 teaspoons baking powder

1 teaspoon bicarbonate of soda
1 teaspoon salt
225 g (8 oz) grated cheese
1 egg
420 ml (14 fl oz) buttermilk or semi-skimmed milk and lemon juice
Oatmeal or rolled oats
2 teaspoons sugar

Method:

1. Preheat the oven to 190°C (375°F, gas mark 5). If you are using milk, stand it in a jug for about 15 minutes with the juice of a lemon to thicken.
2. In a large bowl, sift the plain flour, baking powder, bicarbonate of soda and salt together and then add the wholemeal flour, sugar and nearly all the cheese.
3. In another bowl, mix together the egg and buttermilk. Add this to the flour mixture by making a well in the centre of the dry ingredients and gradually incorporating the liquid.
4. Mix until combined and then turn onto a surface that has been sprinkled with the oats and knead well.
5. Shape the dough into a flattened ball and place on a greased baking tray. Cut a deep cross into the top with a floured knife. Bake for about an hour, sprinkling with the rest of the cheese for the last 5 minutes of cooking.

SODA BREAD

This is the easiest bread to make. It uses buttermilk, which gives a distinctive flavour. If, like me, you have trouble getting hold of buttermilk, you can make the milk sour with the addition of lemon juice.

You will need:

250 g (9 oz) plain flour
250 g (9 oz) wholemeal flour
2 teaspoons bicarbonate of soda
½ teaspoon salt
25 g (1 oz) butter, cut in pieces
450 ml (16 fl oz) buttermilk or semi-skimmed milk
Juice of a lemon (if using milk)

Method:

1. Preheat the oven to 220°C (425°F, gas mark 7) and dust a baking sheet with flour.
2. If using milk, pour it into a jug and add the lemon juice. Leave to stand for 15 minutes.
3. Sift the white flour, salt and bicarbonate of soda into a large bowl and add the wholemeal flour.
4. Rub in the butter. Make a well in the centre and gradually pour in the buttermilk or the milk. Combine from the centre with a wooden spoon or your fingers, handling it gently. The dough should be soft but not sloppy. If it gets too wet add a little more flour.
5. Turn onto a floured board and shape it into a flat, round loaf, about 5 cm (2 in) thick.
6. Put the loaf onto the baking sheet and score a deep cross in the top with a floured knife. Bake for 20 to 25 minutes until the bottom of the loaf sounds hollow when tapped. Reduce the heat to 190°C (375°F, gas mark 5) and cook for a further 25 minutes, or until the crust is browned.
7. Transfer to a wire rack and eat while still warm.

CORNBREAD

Cornmeal gives a nutty flavour and an interesting texture to bread. You can make sweeter bread with more honey and some dried fruit, or add cheese and a dash of paprika for a more savoury recipe.

You will need:

175 g (6 oz) cornmeal
1½ teaspoons bicarbonate of soda
½ teaspoon salt
150 ml (5½ fl oz) plain yoghurt
2 teaspoons honey
2 tablespoons oil
1 egg

Method:

1. Preheat the oven to 200°C (400°F, gas mark 6). Prepare a 15 cm (6 in)greased cake tin.
2. Mix the cornmeal with the bicarbonate of soda and the salt.

3. Whisk the egg and add the honey, yoghurt and oil. Beat well.
4. Combine the two mixtures to make a soft batter.
5. Spoon into the tin and bake for 15 to 20 minutes, until firm to the touch.
6. Turn out and cool on a wire rack.

SINGIN' HINNIE

This recipe from Northumberland gets its name from the sound it makes while it is cooking. You need a spare pair of hands, or at least a pair of fish slices, for the turning.

You will need:

350 g (12 oz) plain flour
½ teaspoon bicarbonate of soda
1 teaspoon cream of tartar
75 g (3 oz) butter
125 g (4 oz) currants
200 ml (7 fl oz) milk

Method:

1. Sift the flour, bicarbonate of soda and cream of tartar together into a bowl.
2. Rub in the butter and add the currants.
3. Make a well in the centre and mix in the milk. Knead lightly to form a dough.
4. Turn onto a floured board and make into flattened balls, about 2 cm (¾ in) thick and 25 cm (10 in) across.
5. Grease and heat a griddle or pan over a moderate heat. Put in the dough and cook until browned. Turn over and cook the other side. This should take about 15 minutes.
6. Serve hot, straight from the griddle. Split crossways and butter wedges to enjoy.

DATE AND WALNUT LOAF

You will need:

225 g (8 oz) self-raising flour
225 g (8 oz) chopped dates
25 g (1 oz) margarine
60 g (2 oz) walnut pieces
110 g (4 oz) brown sugar
1 teaspoon bicarbonate of soda
175 ml (6 fl oz) boiling water
Pinch of salt
1 egg

Method:

1. Put the dates into the boiling water and leave to cool.
2. Preheat the oven to 180°C (350°F, gas mark 4) and grease a loaf tin.
3. Sift the flour with the bicarbonate of soda and salt. Rub in the margarine.
4. Stir in the sugar, walnuts and beaten egg. Mix well and then add the dates and water.
5. Bake in the tin for about an hour and cool on a wire tray.

YOGHURT SCONES

These low fat, low sugar scones can be made into savoury versions as well. See below for cheese options.

You will need:

150 ml (5 fl oz) plain, low fat yoghurt
250 g (9 oz) wholemeal flour
½ teaspoon baking powder
½ teaspoon salt
50 g (2 oz) butter or margarine
1 tablespoon brown sugar

You will need for finishing:

Milk for glazing
Chopped nuts or sesame seeds

Method:

1. Preheat the oven to 220°C (425°F, gas mark 7).
2. Put the flour and salt into a large bowl and sift in the baking powder.
3. Rub in the butter or margarine and the sugar and mix with the yoghurt.
4. Turn onto a floured board and knead lightly to a soft dough. Roll to 2 cm (½ in) thick and cut into rounds.
5. Place on a baking tray and glaze with milk and nuts or seeds.
6. Bake for about 15 minutes before cooling on a wire tray.

CHEESE SCONES

Follow the yoghurt scones recipe above, but add to the flour:

You will need:

1 teaspoon mustard powder
Pinch of cayenne pepper
Instead of the sugar, add 80 g (3 oz) grated cheese

BACON AND POTATO SCONES

This is a good way of using up leftover mashed potato . . . if you ever have leftovers. Alternatively, plan ahead and cook extra potatoes the day before.

You will need: **Makes about 10 scones**

50 g (2 oz) butter or margarine
110 g (4 oz) mashed potato
40 g (1½ oz) parmesan cheese, grated
1 tablespoon oil
2 tablespoons milk
Freshly ground black pepper
3 slices back bacon
175 g (6 oz) plain flour
2 teaspoons baking powder
1 tablespoon fresh parsley or thyme, chopped

Method:

1. Preheat the oven to 220°C (425°F, gas mark 7).
2. Chop the bacon finely. Heat the oil in a frying pan and fry the bacon until crispy.
3. Meanwhile, sift the flour and baking powder into a large bowl. Add the butter and rub in until the mixture resembles fine breadcrumbs.
4. Add the potato, bacon, pepper, cheese, herbs and milk and combine until a soft dough is formed.
5. Roll out onto a lightly floured surface to a thickness of 2 cm (¾ in). Cut into 5 cm (2 in) rounds. Brush with a little milk and bake for about 10 minutes, until golden.

HERB AND WALNUT SCONES

These go well with soup or salad. You can add a little grated cheese to the tops, if you like.

You will need: **Makes about 10 scones**

- 350 g (12 oz) wholemeal flour (or substitute half white flour)
- ½ teaspoon salt
- 4 teaspoons baking powder
- 50 g (2 oz) chopped walnuts or hazelnuts
- 2 tablespoons fresh herbs, chopped (rosemary, parsley or mixed)
- 50 ml (2 fl oz) olive oil (or vegetable oil)
- 1 egg
- 200 ml (7 fl oz) semi-skimmed milk

Method:

1. Preheat the oven to 220°C (425°F, gas mark 7).
2. Put the flour, salt and baking powder in a bowl and stir in the nuts and herbs.
3. In a separate bowl, combine the egg, oil and milk.
4. Add to the dry ingredients and mix to form a dough (don't handle too much).
5. Roll to a thickness of 2 cm (¾ in) on a floured surface and cut into 5 cm (2 in) rounds with a cutter or drinking glass. Glaze with a little milk.
6. Bake for about 10 minutes and serve warm.

TREACLE SCONES

You will need:

450 g (1 lb) plain flour
½ teaspoon bicarbonate of soda
1 teaspoon cream of tartar
½ teaspoon mixed spice
½ teaspoon ground cinnamon
1 tablespoon black treacle
Pinch of salt
1 teaspoon sugar
150 ml (5 fl oz) milk
25 g (1 oz) margarine or butter

Method:

1. Preheat the oven to 220°C (425°F, gas mark 7) and lightly grease a baking tray.
2. Sift together all of the dry ingredients.
3. Rub in the margarine or butter and add the treacle and sugar.
4. Pour in nearly all the milk and mix to a soft dough, adding more milk if necessary.
5. Knead lightly on a floured surface and roll to a thickness of about 2 cm (1 in). Cut into triangles, brush with milk and bake for 12 to 15 minutes.

OATCAKES

These simple oatcakes are easy to make and very good for you. They can be served with butter, low fat cream cheese or hummus. If you prefer, use a cutter to make individual oatcakes.

You will need:

350 g (12 oz) fine oatmeal
1 teaspoon salt
150 ml (¼ pint) boiling water
40 g (1½ oz) soft margarine
Pinch of bicarbonate of soda

Method:

1. Preheat the oven to 150°C (300°F), gas mark 2.

2. Put the oats, salt and bicarbonate of soda into a large bowl.
3. Melt the margarine in water over a low heat. Add to the oats and mix to a dough.
4. Turn out onto sprinkled oatmeal and knead until you have a smooth dough.
5. Dust a rolling pin with oatmeal, halve the dough and roll out into two 25 cm (10 in) rounds.
6. Cut each into 8 sections and place onto greased baking sheets.
7. Bake in a cool oven for 1 hour until crisp. Cool on a wire rack.

PIZZA

This pizza base is yeast-free and quick and easy to make. Children love making and decorating their own versions and they can be made with a view to healthy eating. You could go all out and create faces with the topping for parties or just for fun. This amount will make 6 mini pizzas.

You will need for the base:

80 g (3 oz) self-raising flour
80 g (3 oz) wholemeal flour
1 teaspoon baking powder
Salt and pepper
110 ml (4 fl oz) milk
50 g (2 oz) grated cheese (optional)

You will need for the topping:

4 to 5 tablespoons tomato paste or relish
6 fresh tomatoes, sliced
175 g (6 oz) mozzarella, sliced

Plus some of the following to decorate (think of the eyeball effects and mouths you can make from some of these!). The possibilities are endless:

Green pepper, sliced thinly for hair, or whiskers
6 mushrooms, sliced
Stoned olives
Salami sausage slices

Method:

1. Sift the flour, baking powder and salt together into a large bowl. Add the wholemeal flour and pepper and the cheese, if using. Make into a soft dough with as much of the milk as you need.
2. Preheat the oven to 200°C (400°F, gas mark 6) and grease some baking sheets.
3. Roll the dough out onto a floured surface and cut into 10 cm (4 in) circles or shapes of your choice.
4. Put some of the tomato paste or relish onto each base, followed by sliced tomatoes and mozzarella cheese. Top with your chosen vegetables to create faces or whatever grabs you.
5. Bake for 7 to 8 minutes, or until bubbling and golden, but still recognisable.
6. Allow to cool slightly. Mind the hot cheese!

Fun Activities with Bicarbonate of Soda

As bicarbonate of soda has many properties and is a fairly harmless substance to handle, it is a good compound for explaining some scientific phenomena to children. You don't have to be a child, or a parent or grandparent, to read this section, but it will probably help to have some children to share the activities with. One of the good things about being a primary school teacher was that I had special licence to play, even if my own kids weren't interested.

Most of the following activities are suitable for Key Stage 2 (age 7 to 11) and over.

The activities, while harmless enough, should be overseen by a grown up . . . well, by someone who is grown up in years, anyway – there's no hope for some people! Where safety advice is suggested, please take it, and be prepared to clean up the mess afterwards. As well as providing entertainment, the activities give some good, practical explanations of basic chemistry, something that was sadly lacking in my own school days at an all girls' grammar school. Happily, school science is a lot more fun nowadays.

LEMONADE

This is something most Victorian children would have known about and enjoyed as a novelty. The results are nothing like the sticky, carbonated drinks of today, and quickly lose their fizz, but they are fun to make. When we took classes of children to an Edwardian mansion/ museum in Brighton (Preston Manor) for a role-play history lesson the children learned, among other useful things, how to use bicarbonate of soda to clean the silver, to make gingerbread and finally, to make lemonade.
It was the quietest two hours ever, as the children were being 'interviewed' for jobs as maids or odd-job men by professional actors. The children were not allowed to speak unless spoken to by the 'Housekeeper' or 'Cook', but the look on the children's faces was priceless when the lemonade was made as they watched.

HOME-MADE LEMONADE

This is how children were given a treat in hot weather in the days before bottles of fizzy drinks were available everywhere. You need to be ready to drink it immediately, as the fizz doesn't last long.

You will need:

1 lemon
2 tablespoons icing sugar
1 teaspoon bicarbonate of soda
500 ml (¾ pt) cold water
Drinking glasses
Large jug
Orange or lemon juicer

Method:

1. Make sure you have everything you need before you start.
2. Cut the lemon in half and squeeze out the juice on the juicer.
3. Put the icing sugar into the jug with the bicarbonate of soda.
4. Pour in the water and stir well. This will make the

solution cloudy, but most of the sugar should dissolve.
5. Quickly pour the lemon juice into the jug and stir again. As soon as the fizzing dies down, pour into glasses and enjoy.

GRANDMOTHER'S LEMONADE CORDIAL

For the more discerning palate you might like to try this recipe, which makes a flat cordial for dilution. This is not really a recipe for children to make unattended. The cordial will keep for a couple of weeks in the fridge.
If you like, you can dilute the cordial with soda water, to make it fizzy.

You will need:

6 lemons
1 kg (2 lb) white sugar
25 g (1 oz) citric acid
1.4 l (3 pt) water (boiling)

Method:

1. Grate the rind from three lemons. Squeeze the juice from all six lemons.
2. In a large, heatproof bowl or saucepan dissolve the sugar and citric acid in the boiling water. Stir in the lemon rind and juice.
3. Allow to cool before bottling or storing in a covered jug.
4. To drink, dilute one part of the cordial in four parts water.

MAKING OTHER FIZZY DRINKS

You can make your own special fizz powder which, when added to drinks, makes them effervesce. Don't drink too much of it in one day, though, as it may not be too good for your digestion.

You will need for the powder:

6 teaspoons citric acid
3 teaspoons bicarbonate of soda
2 tablespoons icing sugar

Method:

1. Mix the citric acid crystals to the bicarbonate of soda in a bowl. Crush the mixture with a spoon to form a fine powder.
2. Add the icing sugar, mix thoroughly and put into a dry, clean, labelled jar.
3. To serve, put two teaspoons of powder into a glass and fill it up with the juice drink of your choice, e.g. orange or blackcurrant.

The citric acid reacts with the carbonate in bicarbonate of soda to form carbon dioxide gas, in the same way that vinegar reacts. These bubbles of carbon dioxide are what make your drink fizzy.

FURTHER INVESTIGATION

Try adding half a teaspoon of bicarbonate of soda to half a glass of different drinks, e.g. grapefruit juice, orange squash, blackcurrant cordial. The fizziest drink is the most acidic one. Don't drink all these samples, as there is quite a lot of bicarbonate of soda involved.
Make coloured drinks by adding a few drops of food colouring to water and mixing with your powder, e.g. green.

HOW TO MAKE A SHERBET DIP

Sherbet dips were one of the treats I really loved as a child. They didn't seem to change over the years like some confectionery and I remember the yellow tube with a liquorice straw poking out of the top. They didn't seem to change in price much either, but although the container stayed the same, the contents seemed to steadily decrease in amount.

You could have a lot of fun designing and making personalised bags, tubes or cones, or decorating plastic food bags for party guests. You can present them each with a dipper to take home. Keep the amounts small, as you may get a sore mouth from the citric acid if you overdo it.

You will need for 6 dips:

50 g (2 oz) icing sugar
¾ teaspoon bicarbonate of soda
¾ teaspoon citric acid
6 lollipops or liquorice sticks to dip

Method:

1. Mix all the powders together thoroughly, divide into separate containers, and put a lolly or liquorice stick into each.

When you put the sherbet on your tongue, the citric acid crystals dissolve and react with the bicarbonate of soda and the saliva in your mouth. Citric acid is the same acid as that in lemons and bicarbonate of soda is an alkali. This produces bubbles of carbon dioxide gas, which cause the tingly feeling on your tongue.

For a similar chemical reaction, but in the bath, take a look at the bath bombs in the personal uses section on page 60.

VOLCANO SIMULATION

Have fun with a simple chemical reaction, showing how bicarbonate of soda reacts with vinegar to form bubbles of carbon dioxide, just like when you mix the two ingredients for cleaning purposes.

You will need:

A large dish, washing up bowl or tray
Vinegar
Red food colouring (to simulate lava)
Bicarbonate of soda
A plastic bottle (to simulate the neck of a volcano)
Some sand or gravel

This is best done outside because it can be a bit smelly and messy. If working indoors, put down some newspaper to catch any spills.

Method:

1. First put the newspaper down to protect any surfaces from sand and vinegar.
2. Mix a little red food colouring with some vinegar. This will add a bit of drama to the proceedings.
3. Half fill the plastic bottle with bicarbonate of soda and then stand this bottle in the middle of the dish or tray. Pile sand or gravel around the bottle to form a cone, leaving only the top of the bottle uncovered. If the sand is too dry, wet it a little with some water.
4. Pour the red vinegar into the bottle and stand back and watch the eruption.

The amount of gas produced will depend on the quantities of bicarbonate of soda and vinegar used, so you can experiment with the proportions. The 'eruption' will only last as long as the gas is released. You can top up the bottle with a funnel so that you don't have to rebuild your volcano every time, but the reaction will work best with dry powder and a clean bottle.

BLOWING UP A BALLOON WITHOUT TOUCHING IT

This can prove a bit tricky for small fingers, so help will be needed to assemble the parts. It could form part of a 'magic' act, or challenge for the grown-ups to work out before the act starts. Any vinegar will do, but if you use white vinegar, it will look like ordinary water and get people guessing! If you are out to impress your audience, practise the technique first like all good teachers do. Once again, you can refine the trick with different quantities of vinegar and bicarbonate of soda.

You will need:

- Vinegar
- Bicarbonate of soda
- An empty, clean plastic bottle, without the lid
- A balloon
- Elastic band

Method:

1. Pour about 20 ml of vinegar into the plastic bottle.
2. Put 2 teaspoons of baking soda into the balloon. Twist the neck of the balloon to stop any bicarbonate of soda getting out before you are ready.
3. Fit the balloon over the top of the bottle, making sure the bicarbonate of soda stays at the bottom of the balloon. You might need an elastic band to secure the balloon with if there is a gap. There needs to be a tight fit so that no air can get in or escape.
4. Lift the balloon up so that the baking soda is tipped into the bottle. Sit back and watch the balloon inflate!

WHAT MAKES THE BALLOON INFLATE?

When you tip the bicarbonate of soda out of the balloon into the bottle, it mixes with the vinegar and a chemical reaction produces carbon dioxide gas. As the gas can't escape, the balloon inflates. This is a really good way of explaining how solids and liquids can form gases. The gas can actually be seen to inflate the balloon as it expands. Gases are something that a lot of children find hard to understand, unless they smell them of course!

ROCKET SCIENCE

This can be as sophisticated as you like, ranging from a simple empty film canister to a proper paper rocket-shaped model blasting off. It must be carried out in the open, as there will be a mess. This activity is probably better suited to older children, who have already made volcanoes and inflated balloons. You can use citric acid instead of vinegar, and it won't be so smelly. It also gives you more time to get organised, as the reaction won't start until you add a few drops of water. Please follow the instructions carefully to avoid eye injuries.

You will need:

A 35 mm plastic film canister with lid
Bicarbonate of soda
Vinegar or citric acid crystals and water
Paper and scissors (optional)
Masking tape (optional)

Method:

1. Put a small amount of bicarbonate of soda (¼ teaspoon) and a few drops of vinegar into the film canister and cover it quickly, before gas is released.
2. Place the canister lid down on the ground. (The lid is the bottom end of your rocket.)
3. Stand back and get ready for blast-off.

Warning: don't return to the canister before it pops! If it fails to go off (as it does sometimes, if there's a small leak around the lid) or seems to have failed, wait for a couple of minutes and then open it very carefully, keeping your face well away.

Rockets function by generating huge volumes of gas in a short time. The release of carbon dioxide provides the thrust of the model rocket engine. As the vinegar and bicarbonate of soda mix, the pressure of the carbon dioxide gas created builds up until the container cannot hold the gas and the pressure is released by forcing the lid off the container. This expulsion of exhaust creates thrust to send the model rocket off the ground.

FURTHER INVESTIGATIONS

You can improve your rocket by altering the amounts of vinegar and bicarbonate of soda. You can also improve it by adding a cone-shaped nose and fins made of paper to the canister.

You can time how long it takes for the lid to come off and experiment with quantities, mixing and so forth, until you can get the lid to come off after a specific time, so that you can do a proper countdown.

Mix the bicarbonate of soda with water to form a paste that will stick inside the lid of the film canister, even when you hold it upside down and put it on the canister. Put the vinegar in the canister and keep it upright as you put the lid on to keep the vinegar and bicarbonate of soda away from each other. When you are ready for launch, turn the canister upside down and place it on the launch site.

CABBAGE INDICATORS

This activity can be used to determine the pH of your water supply. You need red cabbage for this to work. It also shows you how acids and alkalis react with an indicator, in this case, red cabbage.

You will need:

Red cabbage leaves
Very hot water
White vinegar
Bicarbonate of soda
Lemon juice

Method:

1. Put the cabbage leaves in a bowl and pour the hot water over them. Stir them and leave for a few minutes. Alternatively use some drained cabbage water when cooking.
2. Strain the water into a jar and put the cabbage to one side. This should give you a coloured liquid, probably pinky red. The colour will depend on where you live. If you live in a hard water area it will be a bluey purple (very alkaline). If it is red you are in a soft water area.
3. Pour a little lemon juice into the red cabbage juice and give it a shake. See what happens to the cabbage water. If it was blue to start with it should turn red. If reddish it should be even redder.
4. Now add some bicarbonate of soda and see what happens. It should change back again.
5. Now try vinegar. You should be able to start working out which substances are acid and which alkaline by now. Acids turn red and alkalis turn blue to green.

Tap water is rarely neutral, although you might expect it to be. It contains chemicals and impurities which make it slightly alkaline. For truly neutral water you need to use distilled water.

FURTHER INVESTIGATIONS

If you use different jars for each addition of acid or alkali, you will have made your own indicator.
You might like to try some different substances and

predict whether they are acid or alkaline, such as

- Cream of tartar
- Washing soda (sodium carbonate)
- Orange juice
- Sour milk
- Milk of Magnesia

Try the same experiment with elderberries or blackberries, as seasonal variations, instead of red cabbage.

GROWING CRYSTALS ON STRING

A variety of crystals can be 'grown' using common substances. Bicarbonate of soda is a good one to try, as it won't harm anyone and is easy to come by. The activity isn't one that gives instant results, however, so it needs to be planned ahead, or perhaps carried out and left alone for a few days. It shows children how solutions form, and how evaporation can be used to retrieve solids from liquids.

You will need:

Bicarbonate of soda
Very hot water
Jam jar
Piece of card to cover the top of the jar
String or thread
Sticky tape

Method:

1. Pour the water into the jam jar and stir in as much bicarbonate of soda as will dissolve.*
2. Make a hole in the centre of the card and thread through the string. Secure with a knot or some tape and suspend the string in the solution, with the card placed on top.
3. Swish the jar gently after about 15 minutes and then again after another 15 minutes. Secure the card with some sticky tape and put the jar somewhere safe, where it won't be disturbed.

*You can pour a little of the solution into a saucer and place in a warm, safe place to speed up evaporation for smaller or less patient children.

The crystals should start to grow after an hour or so, but will take a lot longer to become a feature.

FURTHER INVESTIGATIONS

Try other household substances, such as

- Salt (sodium chloride)
- Sugar
- Washing soda (sodium carbonate)
- Bicarbonate of soda and cream of tartar

This method uses one part bicarbonate of soda to three parts cream of tartar. Mix the bicarbonate in 100 ml (4 fl oz) water. Add one teaspoon of cream of tartar at a time, as the substances will react with each other. Stir until the bubbles have gone and then continue as above.

You can start a small crystal off by the saucer method, then carefully tie this to a thread and suspend in the solution, as above. This is a more secure method for growing much bigger crystals.

APPLE SCULPTING

This activity shows how you can extract Adam's Ale from an apple (but not Adam's apple!).

You will need:

An apple
Bowl
Salt
Bicarbonate of soda

Method:

1. Peel an apple and carve eyes, a nose and a mouth into it. You can weigh it if you want to see how much change takes place.
2. Place the apple in a small bowl and completely cover it with a mixture of half salt and half bicarbonate of soda.

Replace this whenever the salt/bicarbonate gets soggy. Leave for a few weeks and weigh again.

After a while the apple will have become completely desiccated. It will stay like this for a long time, so find a place to display it.

The bicarbonate of soda and salt remove water from the fruit by a process called osmosis. Water passes across the membranes in the cells of the apple to the dry powder. Here, there is a very low concentration of water and over time the water is completely removed. The dry apple will keep without decaying because bacteria and fungus need moisture to grow and the high salt levels will prevent anything from attacking it. If you weighed the apple you will see how much of it was made up of water.

CARTESIAN DIVER OR SUBMARINE

I wonder how many people remember getting divers or submarines in their cornflake packets as a child. I used to have to fight my brother for them, but being younger and not at all scientifically minded, I didn't really know what to do next, and finding a bottle with a wide enough neck to put the diver in wasn't that easy, either. Half the time they didn't work properly. I've now discovered how they work and that there are people all over the world who are still fascinated by them. You can even buy them on-line! The toy divers, frogmen, fish and submarines were placed inside large bottles of water. A small amount of baking powder was loaded into the small plastic model. Baking powder contains two other ingredients as well as bicarbonate of soda, which create acid solutions and react with water to produce carbon dioxide.
Initially the diver or submarine was denser than water, so sank to the bottom, but as the chemical reaction took place, the bubbles of carbon dioxide made the diver less dense, so it rose to the top. When the diver rose to the surface gas escaped, making the diver sink again.
This activity is nearly as much fun, but I wouldn't mind having a go with the originals again!

You will need:

2 litre drink bottles with caps
Small bowl or glass
Medicine dropper

or

A paper clip and half a plastic drinking straw
Plasticine or blu-tack

Method:

1. Completely fill the bottle with water and fill the bowl as well.
2. Squeeze the rubber end of the medicine dropper while it is in the water. You need just enough water in the dropper to let it float in the bowl. If you haven't got a dropper, make a model with the straw bent in half and secured with the paper clip to trap some air in. Use just enough plasticine to make the vessel float on the surface.
3. Place the dropper or paper clip model in the bottle and screw on the cap.
4. Gently squeeze the bottle and the dropper will sink to the bottom. If you stop squeezing the dropper will rise to the top again.

Obviously this is not exactly the same, because there is no bicarbonate of soda or baking powder involved, but hopefully it will inspire greater experiments into finding a substitute for the cornflake divers and submarines.

'LAVA LAMP' POPCORN

This uses the same principle as the divers, but with bicarbonate of soda and vinegar.

You will need:

A tall drinking glass
Water
Bicarbonate of soda
White vinegar
Popping corn that hasn't been popped

Method:

1. Fill the glass about two-thirds full of cold water.

2. Add a heaped teaspoon of bicarbonate of soda and stir.
3. Add some vinegar; just enough so that the water starts to bubble.
4. Quickly add a few pieces of corn, and watch carefully.

The bubbles produced stick to the corn, making it float up to the top. When the bubbles reach the top of the glass they burst, so the corn sinks back down again. Then more bubbles form and stick to it and up it goes again. As long as the bicarbonate of soda and vinegar keep making gas, your corn keeps on rising and falling.

FURTHER INVESTIGATIONS
Try different objects in the glass.

Appendix 1 — cleaning methods

CLEANING PROBLEM	CLEANING SOLUTION
Floor cleaner (for vinyl and tiles)	Dissolve 4 tablespoons bicarbonate of soda in a bucket of warm water.
Hob cleaner	Paste of 4 tablespoons bicarbonate of soda, squeeze of lemon juice, warm water and a teaspoon of vinegar.
Oven cleaner	Equal parts of salt and bicarbonate of soda, mixed with vinegar.
Burnt-on food 1	Pour a layer of bicarbonate of soda onto the base of the pan and use just enough water to moisten the soda. Leave until the next day, then scrub clean.
Burnt-on food 2	Mix 1 spoonful of salt with 2 of bicarbonate of soda with a little lemon juice. Spread this over the offending article and leave to dry. Rub off with a scouring pad.
Heavily encrusted pans	Put a layer of water in the bottom and add a cup of vinegar. Bring this to the boil, remove from the heat and then sprinkle in 2 tablespoons of bicarbonate of soda.
Brass polish 1	2 tablespoons of flour, 1 tablespoon of salt and 1 tablespoon of bicarbonate of soda, mixed together with vinegar to make a thick paste.
Brass polish 2	Sprinkle some bicarbonate of soda onto half a lemon and rub clean.
Silver polish 1	Thin paste of bicarbonate of soda and water. Spread over the object and let it dry. Polish with a soft, clean cloth.
Silver polish 2 (heavily tarnished)	Fill a plastic bucket with hot water and put a piece of aluminium foil in the bottom. Sprinkle bicarbonate of soda over the item for cleaning and put into the water. Leave for 15 minutes then dry with a soft cloth. Smaller items can be washed in an old aluminium foil baking tray in the same way.
Tile and grout cleaner	2 parts bicarbonate of soda to 1 part white vinegar, applied with an old toothbrush.
Bath and basin cleaner	2 parts bicarbonate of soda to 1 part vinegar. Apply with a damp cloth. Rinse with cold water.
Shower curtains	Put them in the bath and brush or sponge with 4 or 5 tablespoons of bicarbonate of soda. Rinse with hot water and white vinegar. Dry thoroughly.
Spot carpet stain remover	Dab with a cloth and some soda water immediately.
Carpet cleaner	Use bicarbonate of soda to soak up as much of the liquid as possible. This can then be swept up and the carpet dabbed with a mild solution of bicarbonate of soda and warm water. Allow to dry thoroughly before vacuuming.
Drains 1	Equal quantities of salt and bicarbonate of soda powder, followed by hot water.
Drains 2	Use half a tub of bicarbonate of soda and a cup of vinegar. Leave this for an hour then pour down lots of very hot water.
Bath cleaner 2 (very dirty)	1 part bicarbonate of soda, 4 parts vinegar and a squeeze of lemon juice.
Air freshener	A few drops of fragrance oil in half a tub of bicarbonate of soda. Leave open on an out of the way shelf.